Improving traceability in food processing and distribution

Related titles:

Handbook of hygiene control in the food industry
(ISBN-13: 978-1-85573-957-4; ISBN-10: 1-85573-957-7)
The foundation of food safety lies in good hygiene practice. With its distinguished editors and over 40 contributors, this comprehensive reference will set the standard for good hygiene management. It discusses both good hygienic design, cleaning and monitoring techniques.

Detecting allergens in food
(ISBN-13: 978-1-85573-728-0; ISBN-10: 1-85573-728-0)
Allergens pose a serious risk to consumers, making detection a major priority for food manufacturers. Bringing together key experts in the field, this important collection reviews the range of analytical techniques available and their use to detect specific allergens.

Food authenticity and traceability
(ISBN-13: 978-1-85573-526-2; ISBN-10: 1-85573-526-1)
The need to trace and authenticate food products and ingredients has never been more important. The first part of this book reviews established and emerging authentication techniques. Part II discusses how they are applied to particular foods. The final part of the book looks at current work on traceability systems in the food industry.

Details of these books and a complete list of Woodhead's food science, technology and nutrition titles can be obtained by:

- visiting our web site at www.woodheadpublishing.com
- contacting Customer Services (e-mail: sales@woodhead-publishing.com; fax: +44 (0) 1223 893694; tel.: +44 (0)1223 891358 ext. 30; address: Woodhead Publishing Ltd, Abington Hall, Abington, Cambridge CB1 6AH, England)

If you would like to receive information on forthcoming titles, please send your address details to: Francis Dodds (address, tel. and fax as above; e-mail: francisd@woodhead-publishing.com). Please confirm which subject areas you are interested in.

Improving traceability in food processing and distribution

Edited by
Ian Smith and Anthony Furness

CRC Press
Boca Raton Boston New York Washington, DC

WOODHEAD PUBLISHING LIMITED
Cambridge England

Published by Woodhead Publishing Limited, Abington Hall, Abington,
Cambridge CB1 6AH, England
www.woodheadpublishing.com

Published in North America by CRC Press LLC, 6000 Broken Sound Parkway, NW,
Suite 300, Boca Raton, FL 33487, USA

First published 2006, Woodhead Publishing Limited and CRC Press LLC
© 2006, Woodhead Publishing Limited, except Chapter 11 which is
© 2006 Anthony Furness and AIM UK
The authors have asserted their moral rights.

British Library Cataloguing in Publication Data
A catalogue record for this book is available from the British Library.

Library of Congress Cataloging in Publication Data
A catalog record for this book is available from the Library of Congress.

Woodhead Publishing ISBN-13: 978-1-85573-959-8 (book)
Woodhead Publishing ISBN-10: 1-85573-959-3 (book)
Woodhead Publishing ISBN-13: 978-1-84569-123-3 (e-book)
Woodhead Publishing ISBN-10: 1-84569-123-7 (e-book)
CRC Press ISBN-10: 0-8493-9160-1
CRC Press order number: WP9160

The publishers' policy is to use permanent paper from mills that operate a
sustainable forestry policy, and which has been manufactured from pulp
which is processed using acid-free and elementary chlorine-free practices.
Furthermore, the publishers ensure that the text paper and cover board used
have met acceptable environmental accreditation standards.

Typeset in India by Replika Press Pvt. Ltd.
Printed by TJ International, Padstow, Cornwall, England

Contents

Contributor contact details

(* = main point of contact)

Editors

Mr Ian G. Smith
AIM UK
The Old Vicarage
All Souls Road
Halifax HX3 6DR
UK

E-mail: ian@aimuk.org

Professor A. Furness
AIM UK
The Old Vicarage
All Souls Road
Halifax HX3 6DR
UK

E-mail:
Anthony@furness5079.fsnet.co.uk

Chapter 1

Professor K. A. Osman
Centre for Automatic Identification
and Intelligent Systems
Technology Innovation Centre
Millennium Point
Birmingham B4 7XG
UK

E-mail: keith.osman@tic.ac.uk

Professor A. Furness
AIM UK

E-mail:
Anthony@furness5079.fsnet.co.uk

Chapter 2

Dr Floor Verdenius
Agrotechnology & Food
Innovations
Wageningen University and
Research Centre
Bornsesteeg 59
6708 PD Wageningen
The Netherlands

E-mail: Floor.Verdenius@wur.nl

Chapter 3

Dr Frans-Peter Scheer
Packaging, Transport and Logistics
Agrotechnology & Food
Innovations
Wageningen University and
Research Centre
Bornsesteeg 59
6708 PD Wageningen
The Netherlands

E-mail: Frans-Peter.Scheer@wur.nl

Chapter 4

Drs Ing Lars Hulzebos* and
Drs MTD Nicole Koenderink
Agrotechnology & Food
Innovations
Wageningen University and
Research Centre
Bornsesteeg 59
6708 PD Wageningen
The Netherlands

E-mail: Lars.Hulzebos@wur.nl
E-mail: Nicole.Koenderink@wur.nl

Chapter 5

Drs Nicole Koenderink* and
Drs Lars Hulzebos
Agrotechnology & Food
Innovations
Wageningen University and
Research Centre
Bornsesteeg 59
6708 PD Wageningen
The Netherlands

E-mail: Nicole.Koenderink@wur.nl
E-mail: Lars.Hulzebos@wur.nl

Chapter 6

Mr Michael Klafft*
Institute of Information Systems
Humboldt University, Berlin,
Spandauer Str. 1 10178 Berlin
Germany

Mr Julien Huen, Mr Claus Kuhn,
Mrs Elsa Huen and
Mr Stefan Wößner
Fraunhofer Institut für
Produktionstechnik und
Automatisierung
Nobelstraße 12
70569 Stuttgart
Germany

E-mail: jbh@ipa.fhg.de or
M.Klafft@wiwi.hu-berlin.de

Chapter 7

Professor Maria F. Camões*
Department of Chemistry and
Biochemistry
Faculty of Sciences
University of Lisbon
Campo Grande
Building C8
P-1749-016 Lisbon
Portugal

E-mail: mfcamoes@fc.ul.pt

Dr R. Bettencourt da Silva
Directorate General for Crop
Protection
Quinta do Marquês
P-2780-155 Oeíras
Portugal

E-mail: ricardosilva@dgpc.min-
agricultura.pt

Chapter 8
Dr J. A. Lenstra
Faculty of Veterinary Medicine
University of Utrecht
PO Box 80154
3508 TD Utrecht
The Netherlands

E-mail: j.a.lenstra@vet.uu.nl

Chapter 9
Dr Andre Poucet
IPSC
European Union Joint Research
Centre
Via Enrico Fermi 1
21020 Ispra (VA)
Italy

E-mail: andre.poucet@cec.eu.int

Chapter 10
Dr Raoul Vernède* and Dr Ingrid
Wienk
Agrotechnology & Food
Innovations
Wageningen University and
Research Centre
Bornsesteeg 59
6708 PD Wageningen
The Netherlands

E-mail: Raoul.Vernede@wur.nl
E-mail: Ingrid.Wienk@wur.nl

Chapter 11
Professor A. Furness
AIM UK

E-mail:
Anthony@furness5079.fsnet.co.uk

Part I

Traceability, safety and quality

1

Developing traceability systems across the food supply chain: an overview

A. Furness, AIM UK and K. A. Osman, Centre for Automatic Identification and Intelligent Systems, UK

1.1   Introduction

Traceability is now emerging as a 'watch-word' for consumer and regulatory confidence with respect to food quality, food safety and the infrastructure for producing, processing and delivering food products from the point of origin to the point of sale. Various definitions have been derived for traceability, including a European Union (EU) General Food Law Regulation definition in which traceability is defined as: 'the ability to trace and follow a food, feed, food-producing animal or substance through all stages of production and distribution' (EU Regulation Food Law: 8/5/01). An International Standards Organisation (ISO) definition is also to be found that defines traceability as: 'the ability to trace the history, application or location of an entity by means of recorded information' (ISO 8402:1994), The ISO definition, whilst more general in terms of the traceable entity, draws attention to the importance of recorded information that is essential for satisfying traceability requirements.

The need for traceability has arisen primarily from consumer and government concerns over food safety, hygiene and authenticity. These concerns range from the human risks associated with animal-borne pathogenic organisms, such as salmonella, listeria, clostridium and E-coli O157 through agents causing bovine spongiform encephalitis (BSE) to those risks perceived for genetically modified (GM) vegetables, cereals, fruits and GM influenced foods in general. As a consequence of national needs and sectoral responses for traceability, sensitised by developments in global trade and consumer demands, a number of traceability guidelines have been developed. Notably these guidelines include the Traceability of Fish – Application of EAN.UCC Standards (EAN International); Traceability of Beef Guidelines (EMEG);

Fresh Produce Traceability Guidelines (EAN International); Traceability in the Supply Chain (GENCOD EAN France) and Traceability Implementation (EAN.UCC project). In response to initiatives such as these and to the burgeoning requirements for legislative compliance, increasing numbers of representative bodies for food sector supply chains are deriving and implementing traceability systems. With varying degrees of supply chain coverage and an emphasis on particular foodstuffs such as fish, meat and wine, such systems can satisfy some traceability requirements for these products but cannot readily be expanded to encompass other products. As an increasing number of supply chains attempt to develop sector-specific traceability systems, problems will inevitably occur where the ability to connect across different supply chain boundaries will be essential to provide traceability of multi-ingredient food products. A fruit pie, for example, will have a multiplicity of ingredients stemming from many different supply chains.

Complexities will arise with respect to the range and variation in the accessible information required for traceability. They will also arise with respect to incompatibilities in systems' structures for transferring and accommodating appropriate information. While food safety and food quality may be seen as the primary drivers for traceability, other needs can and will be recognised for which traceability systems will be required. Food authentication, quality assurance, label verification, shrinkage (product loss) management, process development and consumer support are further areas in which traceability can be seen to be necessary, each characterised by particular functional needs that will increase overall complexity. This also draws attention to a system requirement for being able to distinguish and satisfy particular traceability functions using a common traceability system structure – separating traceability functions from system requirements.

It has to be recognised that the food production and supply industry is highly integrated and of global significance in terms of trading reach and trading opportunities. Harmonisation of traceability systems is required in order to avoid the almost inevitable chaos that could ensue through lack of compatibility and effective management of complexity. This becomes increasingly significant as traceability systems are structured to make more and more use of information and communications technology (ICT) and the opportunities for powerful and effective information transfer across communication networks such as the Internet.

Aspects of commonality and moves towards exploiting existing standards for numbering and identification are, fortunately, evident in many of the systems under development. So too are developments in electronic data communications for business support purposes, offering potential for integration into traceability systems. Despite this attention to legacy, the need is still evident for deriving a generic approach to traceability that can handle cross-supply chain interaction and provide a logical framework for harmonisation and systems compatibility. Any approach would also have to support a growing list of traceability functions and allow effective management of supply chain complexity.

Underpinning this generic approach is a deeper understanding of the nature of traceability and the factors influencing the need to implement traceability functions. Firstly, it is important to consider further why traceability is required, including the factors influencing the implementation of traceability systems. Food safety is the primary reason for traceability and various functional components can be distinguished. The UK Food Agency has identified the following functional roles for traceability within the food supply chain:[1]

- **Food safety incidents** – requiring robust traceability to facilitate rapid response to breakdowns in food safety, allowing remedial actions, such as product withdrawals and recalls to be initiated for the purposes of protecting public safety.
- **Food residue surveillance programmes** – using traceability systems to facilitate food sampling at appropriate points throughout the food supply chain, testing for residues, such as pesticides, and mapping to establish where in the supply chain excessive residue levels may have occurred.
- **Risk assessment from food exposure** – where a traceability system can facilitate access to information concerning foods or food ingredients that may have significance with respect to food safety.
- **Enforcement of labelling claims** – using traceability to help resolve allegations of false labelling and to help determine supply chain integrity with respect to food claims.
- **Fraud** – wherein effective traceability, regular audit and reconciliation measures can assist in preventing fraud and theft of food items.
- **Food wastage** – where traceability and associated quality control systems can be applied to speed up and improve food distribution processes and reduce food wastage.
- **Meat hygiene** – where traceability can help enforce and support meat hygiene in processing and handling of food within supply chains.

Other reasons for traceability, often associated with safety, include compliance with food legislation, quality assurance of food producing and handling processes, authentication of food items, process and supply chain development and consumer services in respect of food products. Each of these factors constitutes a traceability function requiring access to relevant, function-defined, item-related information. By suitable partitioning and coding of information into sets, the traceability functions may be more readily accommodated and linked to a common traceability structure.

A significant feature of a traceability system *per se* is that it should be continually available for traceability, yet is used only as and when required. As a stakeholder development requiring capital expenditure and expenditure to maintain, a traceability system may be viewed as an imposition, adding little or no value to the processes concerned. Where then is the incentive to

---

[1]Traceability in the Food Chain, UK Food Agency, Food Chain Strategy Division (Paper Note 02/02/01), 14.02.2002.

implement traceability? Legislation is clearly a driving influence in this respect, but hardly seen as an incentive, unless perceived with respect to consumer confidence. Having to comply with regulatory demands rather forces the issue. However, through appropriate consideration of process structure and functionality, traceability systems may be devised that not only fulfil the traceability requirement, but also add value to the processes concerned. Such developments invariably make use of item-attendant data carriers that allow automatic identification and data gathering and facilitate improvements in process efficiencies and quality. Such developments may also be seen to align with quality practice. The ISO 9000 series of quality standards, ISO 9001 in particular, distinguishes traceability as a requirement for compliance. It requires that a product be traceable from a current stage of existence back through all its stages of manufacture or production by means of either paper-based or computer-based records, accurately and promptly produced for the purpose. By adding value to the processes an incentive-based approach to traceability can be promoted.

1.2   Accommodating multi-functional traceability requirements

In gaining further insight into the generic approach to accommodating multi-function traceability requirements, it is necessary to distinguish further between the traceability system and traceability functions. A traceability system is required to provide an unambiguous, uninterrupted means of physically tracing and tracking an item, and/or its constituent components, through the inter-linking nodes of a supply chain. A node is distinguished as a point in the chain in which the item is handled or processed in some way. To achieve such a system it is necessary for the item or constituent items concerned to be appropriately identified and that the identifiers provide linkage to the relevant item information that is stored remotely. The need for the linkage is to facilitate access to the item-specific information which in turn can be used to satisfy a traceability or process support function.

Traceability systems may be paper-based or structured to exploit the benefits of information and communications technology (ICT). While not losing sight of the need for paper-based alternatives it is logical to structure systems on technology that can provide more and more efficient and speedy traceability support. To achieve a fully integrated, harmonised approach to traceability requires effective identification of all items within supply chains; from raw materials, through product items, packaged product, logistical units to palletised units; with appropriate coded linkage and communication paths to appropriately coded information sources for satisfying the needs of the various traceability functions. Flexibility must be provided through partitioning of information. In some cases information may also accompany the item or items concerned.

In general the quantity of information required to satisfy a traceability need will be too large to be carried on or with the item. Moreover, various items of information may be required to satisfy a range of process support and traceability requirements. Generally, information concerning items will be held in appropriate databases. To accommodate the diversity and intrinsic complexity of the information requirements, information partitioning and item identification needs, it is necessary to determine the areas of commonality that exist with respect to traceability systems and derive a strategy and a framework that can exploit the features concerned.

All traceability systems, irrespective of supply chain items, industry affiliation and functions supported, can be seen to exhibit common structural features:

- **Item identification**, unambiguous and linkable for accommodating processing and handling in the supply chain.
- **Item-attendant and/or item-associated information** appropriate to nodal transforms and transactions and any inter-nodal events that have a bearing upon traceability.
- **Process-based information** relating and linked to items processed or handled in the supply chain.
- **Communication** links to allow access and exchange of information.

Based upon these features of commonality it is possible to construct a generic structure for a traceability system characterised by a vertical item and data flow (up and down the supply chain) facility and a transverse (supply chain inter-nodal) data gathering, data flow and storage facility. The vertical structure would be minimalist in the sense that it would exploit the simplest of item identifiers and data carriers and provide the necessary code identifier transfer and access links to transverse information stores to support the traceability requirements. The transverse structures accommodate the wealth of item information required for traceability and process support purposes, but appropriately coded into information sets to serve the respective traceability or process functions. The next step in defining a generic framework for traceability is to define the means by which the vertical and transverse components can be inter-linked.

1.2.1   Vertical and transverse inter-linking
The essential features of the generic framework, for a single node in a supply chain, depicting the vertical and transverse structure are illustrated in Fig. 1.1. It is significant in respect of the core item and process entities to distinguish essential information (possibly object-orientated) for more efficient access and routing to additional information. By including appropriate data carrier support for items between nodes the facility can also be provided for better control and performance of inter-nodal events. As a consequence, further information may be generated that must be carried with or associated with

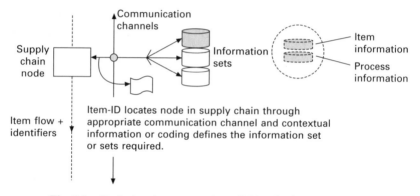

Fig. 1.1   Vertical and transverse inter-linking in the supply chain.

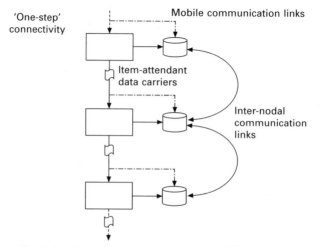

Fig. 1.2   One-step connectivity in a traceability system.

items or communicated directly to appropriate nodal information management systems by mobile communications (Fig. 1.2). These 'one-step' links not only support linkage for traceability purposes, they can also support the flow of information between adjacent nodes for the purposes of added-value process functionality, transaction and supply chain management. The 'one-step' look-forward, look-back structure, appropriately interfaced and extended, may also facilitate first layer, node-to-node ('daisy-chain') communication linking along the supply chain. The information storage facilities and associated information management systems (IMS) may be viewed as traceability control and information support points.

Within a minimalist traceability structure little if any information is conveyed with the item, the principal function of the item-attendant or item-accompanying data carriers being to provide item identification, which is structured and sufficient in form to allow linking and access to information stored in local or linked databases through appropriate communication channels. In some cases the need may be seen for item-attendant, machine-readable, portable data files that allow specific information to accompany the item, the nature and extent of the information depending upon item and nodal-specific needs. For example, a consignment of perishable goods may be accompanied by a data file that acts as a shipping manifest, containing information on both shipment contents and the dispatch time, destination and consignment handling details. This approach ensures that critical information is immediately available with the item as and when required, circumventing problems that might otherwise occur due to breakdown of communication links to remotely held data.

In some cases the amount of information stored in nodal databases could be substantial, particularly where a number of process and traceability functions have to be satisfied. This will almost certainly require the use of large relational databases and, in some instances, linkage with larger, national or other scheme-related databases. This introduces a further layer in the traceability infrastructure. To allow access to these databases and management systems for traceability purposes requires a network infrastructure (including use of the Internet) with appropriately authorised access control and communication protocols (Fig. 1.3).

In defining an infrastructure of this kind for traceability it is necessary to identify standard data structures, identifiers and both interface and communications protocols to meet the needs of different supply chains, with connectivity in and between supply chains and across national boundaries. The identifiers must include coding structures for location, information sets and access control as well as item identifiers.

1.2.2 Location requirements
Being able to identify the nodes, and any other item- and information-handling

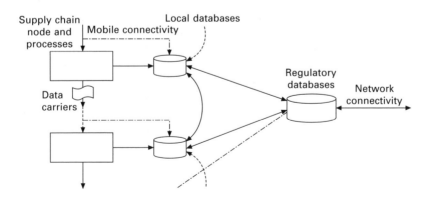

Fig. 1.3 Linking databases in a traceability system.

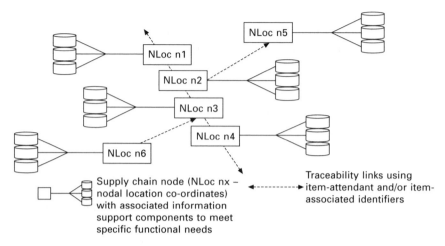

Fig. 1.4   Integrating location requirements into a traceability system.

points in the supply chain, is an essential requirement for engineering a generic, fully integrated traceability system capable of handling appropriately coded and accessible information sets (Fig. 1.4). Various agencies can be recognised for implementing location-based identification, including the EAN.UCC Global Location Number (GLN) and Global Positioning Systems (GPS) defined location codes. Both provide the facility for assigning a code to a particular legal, functional or physical, fixed positional location.

1.3   Item-specific data capture

At an origination node in a supply chain, or where new food items or ingredients enter the chain, it will generally be necessary to establish primary identification of food entities concerned for the purposes of authentication and onward identification. The primary identification techniques are many and varied, and are selected for use depending upon the food entity concerned. Typically techniques include DNA methods, protein and metabolic fingerprinting, NMR spectroscopy and a wide range of analytical techniques for determining food components, as described in other chapters of this book. From a traceability standpoint such techniques provide the means of individually identifying a particular food item on the basis of its intrinsic features, providing a biological 'foot-print' or profile that can be appropriately stored and accessed as required. On this basis any food entity, such as a batch of wheat, an animal, an item of fruit, vegetable or spice, entering the supply chain can be identified individually or by category. Linked by the primary identification there is likely to be further information concerning the item that is held and this forms an accessible information set. To access this information remotely by electronic means may require an additional code or identifier.

Any other identifier used within the supply chain is secondary, and generally involves a 'unique' item, batch or other item-linked number or alpha-numeric string being assigned and attached to a selected item for the purposes of traceability. Ideally, the secondary identifiers should be linked through the assigned coding structure to the primary identifier. In some cases the item-specific, feature-based, primary identifier will be coded to distinguish item type. Legislation is in prospect concerning coded identification of genetically modified organisms (GMO),[2] wherein specific identification codes will be assigned to authorised genetic transformations or modifications to be marketed or used in food products. A further aspect of secondary identification is the assignment of an identification code that can be used to facilitate on-going identification of the item and its handling in subsequent parts of the supply chain. For an ICT supported traceability system it is necessary for the identifier to be machine readable and ideally supported by an appropriate numbering and identification standard (such as the EAN.UCC system, considered later) (Fig. 1.5).

A larger part of the vertical traceability structure will inevitably be achieved using item-attendant identifiers, most probably using standard EAN.UCC numbering and identifier structures. Particular information will relate to items entering and items leaving nodal points along the supply chain. A node must have a link and knowledge of the appropriate item information at one nodal level above and one level below the node concerned in order to satisfy the lowest demand for traceability. The one-step above and one-step below node-to-node links should be seen as constituent links for traceability purposes, with traceability identifiers on item-attendant data carriers being used to achieve linkage. Node-to-node communication links will add to the robustness of linkage for traceability purposes, providing a degree of protection against link failure which would otherwise compromise the traceability chain.

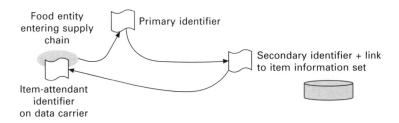

Fig. 1.5   Item-attendant identifiers in a traceability system.

---

[2] Proposal for a Regulation of the European Parliament and the Council concerning traceability and labelling of genetically modified organisms and traceability of food and feed products produced from genetically modified organisms and amending Directive 2001/18/EC.

To achieve the necessary harmonisation and intra-operability of traceability systems will require the use of standardised approaches to item identification, information handling and communications. Fortunately, there exists a legacy of standards that go a long way towards achieving the harmonisation and intra-operability required.

1.4   The EAN.UCC coding system

A significant and widely used system for secondary item-attendant identification is the EAN.UCC System[3] of numbering and identification. The system also extends to identification of locations and services, and has already been applied in practical traceability systems and promoted through guidelines specifically directed at traceability.[4] Six standard numbering structures presently comprise the EAN.UCC System. These structures and their components are discussed below

1.4.1   Global Trade Item Number (GTIN)

Formerly known as the Article Number, the GTIN is used to positively identify trade items. The trade item is recognised as any entity, product or service for which there is a need to retrieve pre-defined item-attendant data at any point within a supply chain. The definition embraces both items and items accommodated in different forms of packaging, for example multi-packs or case level products.

The GTIN itself is defined as a 14-digit number from which a family of four unique numbering structures is derived. The full 14-digit number system, known as the EAN/UCC-14 structure, is used to identify trade items that do not cross a retail point of seller. The structure comprises a leading (left side) 'indicator' digit (or logistical variant) followed by 12 digits for the identification of items. The final digit is a check digit, derived from the other 13 for the purpose of providing a level of error detection. The indicator digit allows each user to increase the numbering capacity when seeking to identify similar trade units accommodated in different packaging configurations. The number 9 in this position is reserved for identifying outer cases of items of continuously variable measure, usually weight.

EAN/UCC-13
The GTIN structure without the indicator digit represents a 13-digit number known as the EAN/UCC-13 structure. This and two other truncated structures

---

[3] EAN – International Numbering Association, UCC – Universal Code Council.
[4] Traceability in the Supply Chain – From Strategy to Practice, GENCOD EAN France (2001).

are used to identify items by type at point-of-sale. The EAN.UCC-13 structure effectively comprises three parts: a company prefix number, an item reference number and a check digit. The company prefix number is made up of an EAN.UCC two or three prefix issued by one of the world-wide network of EAN member organisations to decentralise the administration of identification numbers. Each organisation has a characteristic number: 50 for the UK for example, 00 to 09 for the USA and Canada, 549 EAN-Iceland, 93 EAN-Australia and so forth, covering almost 100 countries. The company part of the number is assigned to the user by the numbering organisation or by the UCC, depending upon whom the user approaches, and can be 7, 8 or 9 digits in length depending upon the user's numbering requirements. The prefix together with the company number form the company prefix number. Within the EAN/UCC-13 structure 12 digits comprise the company prefix number and the item reference. Consequently, the more digits assigned to the company number the fewer are available for item reference, the latter being assigned by the user to distinguish particular items or products. For example:

- If the company number is 5012345, five digits are available for the item reference, allowing 100 000 items to be distinguished.
- If the company number is 50123456, four digits are available for the item reference, allowing 10 000 items to be distinguished.
- If the company number is 501234567, three digits are available for the item reference, allowing 1000 items to be distinguished.

It is the system user's responsibility to ensure that the item reference number is selected to achieve a completed number that is unique for the item type or service being numbered. A check digit forms the final part of the structure and is calculated (automatically within system software) according to a specified algorithm and used to check for errors.

UCC-12 structure
The UCC-12 structure is similar in function and form to that of the EAN-13 structure, comprising a UCC company prefix, item reference and check digit. It is used within the USA and Canada for item type identification at the point of sale.

EAN/UCC-8 structure
The EAN/UCC-8 structure is essentially used for in-store item identification in which a truncated prefix can be used, together with an item reference and a check digit. The numbers so distinguished provide unique identification when processed in a 14-digit data field, zeros being used to fill the leading (left) field positions as appropriate for the EAN/UCC-13, UCC-12 and EAN/UCC-8 structures. It is the field format for GTIN used in all business transactions and EDI messaging supporting the standard.

Although the EAN.UCC system provides a number of structures for identification purposes it is insufficient for achieving unique identification of each and every individual item that crosses the point of sale. Other identifiers are required to achieve this level of granulation. One such scheme is being proposed by the MIT Auto-ID Center and considered by EAN.UCC. Known as the electronic product code (ePC), this scheme is based upon a 96-bit structure, but has still to be endorsed and promoted as a standard for item identification.

1.4.2   Serial Shipping Container Code (SSCC)

The Serial Shipping Container Code is an 18-digit code used to identify logistical units – shipping containers or transport units – the additional digits over the GTIN being to accommodate larger item reference numbers and supplementary information. The leading digit (extreme left), called an extension digit, effectively increases the capacity of the SSCC and is used to qualify the application of the code by assigning values 0–9 in the data field according to specified rules. The remainder of the structure consists of an EAN.UCC company prefix, serial number and a check digit.

1.4.3   Global Location Number (GLN)

The Global Location Number provides a unique numbering system for locations. The Global Location Number (GLN) of EAN.UCC identifies business or organisational entities such as:

- **legal entities**: whole companies, subsidiaries or divisions such as supplier, customer, bank, forwarder and so forth
- **functional entities**: a specific department within a legal entity, e.g., accounting department
- **physical entities**: a particular room in a building, e.g., warehouse or warehouse gate, delivery point, transmission point.

Each location is allocated a world-wide unique identification number. Those GLNs are reference keys for retrieving information from databases such as postal address, region, telephone and fax numbers, contact person, bank account information, delivery requirements or restrictions.

The identification of locations by GLN is required to enable an efficient flow of goods and information between trading partners through EDI messages and other electronic data exchange systems, physical location marking and routeing information on logistic units. The use of GLN provides companies with a method of identifying locations, within and outside their company, that are:

- **unique**: with a simple structure, facilitating processing and transmission of data
- **multi-sectional**: the non-significant characteristic of the EAN.UCC numbers allows any location to be identified and consequently any business regardless of its activity

- **international**: location numbers are unique world-wide. Moreover, the international network of EAN.UCC Numbering Organisation, covering about 100 countries, provides support in the local language.

1.4.4   Global Returnable Asset Identifier (GRAI)
The Global Returnable Asset Identifier is used to identify reusable entities such as containers and totes, which are normally used for the transportation and storage of goods. The structure is essentially 14 digits in length and accommodates the EAN/UCC-13 structure and an optional facility for adding a serial number up to 16 digits in length.

1.4.5   Global Individual Asset Identifier (GIAI)
The Global Individual Asset Identifier is used to identify uniquely an entity that is part of an inventory within a given company. The number, which accommodates the EAN.UCC company prefix, together with a variable length individual asset reference, is up to 30 digits in length.

1.4.6   Global Service Relation Number (GSRN)
The Global Service Relation Number is an 18-digit code comprising an EAN.UCC company prefix and a service reference used to identify the recipient of services from a service provider. It does not therefore identify a particular person or legal entity but a relationship or action that requires an identification point for accommodating transaction data, for example.

1.4.7   Areas of application
These numbering structures effectively service five areas of application, namely the identification of trade items (item type), logistical units, assets, locations and service relations. While each may have value in structuring traceability systems the need remains for a standard, an extension of the EAN.UCC system that can support greater granularity in identifying individual items both within the supply chains and at point of sale. Batch codes are often incorporated into product labels but only in human readable form. For further traceability support the need can be seen for at least machine-readable batch code additions to food labels. Such coding is feasible using low-cost printable composite codes.

1.5   Data carrier technologies

Because of the diversity to be found in supply chains, a generic framework for traceability not only requires a standard approach to numbering and identification, but also requires a range of identifier carrier technologies that can provide flexibility in defining traceability systems. A range of item-attendant technologies can be identified for this purpose. Moreover, technologies may also be identified that can add further functionality and offer opportunities for engineering innovative solutions to supply chain problems. The item-attendant technologies may be grouped as follows:

- **data carrier technologies** – including linear bar codes, two-dimensional (multi-row bar code and matrix codes) and composite codes, contact and non-contact magnetic data carriers, contact memory and radio frequency identification (RFID) data carriers
- **location and locating technologies** – exploiting GLN by carriers including RFID and EANCOM, and active RFID real time locating systems (RTLS) and global positioning systems (GPS) locating technologies
- **communication technologies** – including wireless local area network (WLAN) technologies
- **sensory technologies** – exploiting at the item level developments in sensory and telemetry technologies
- **security technologies** – embracing a range of technologies for fraud prevention and security at packaging level.

Presently the data carrier technologies are the most significant for traceability purposes.

The simplest and most commonly used of the data carrier technologies is the linear bar code, where the data is encoded as a series of narrow and wide black bars and light spaces that can be readily printed onto paper, cardboard, plastic and other substrates by a variety of methods including thermal-transfer, direct-thermal, inkjet and dot-matrix printing. Linear bar codes are generally read by illuminating with a red light source, either by raster scanning a single-spot laser across the symbol or by illuminating with red light emitting diodes and using a charge-coupled device (CCD) detector, demodulating the reflected light to retrieve the modulated response and further decoding this to extract the data. A wide range of scanning devices is available, but all have a 'line-of-sight' constraint, i.e., the bar code must be visible so that it can be illuminated and scanned.

An extension of the linear (1D) bar code is the 2D bar code, in the form of either a multi-row or matrix symbol. PDF417 is a multi-row symbology that is commonly used to encode shipping details, bills of lading and so forth. Such symbols generally facilitate a large data payload in excess of 1,500 characters (Fig. 1.6). The low cost realisation of 2D codes renders them attractive for portable data file applications. For traceability purposes they may be used both to carry identifiers and item-assisted information. As such they can be useful in supply chain processes for real-time labelling and item tracking, particularly where items are being separated or combined with other items (e.g. ingredients for a food product). The printability of 2D code using a variety of printing methods provides considerable flexibility for

EAN-13   PDF-417   DataMatrix EC200

Fig. 1.6 Types of bar code.

tackling in-chain labelling. Using inkjet technology 2D codes can even be printed onto foodstuffs, providing the ink is safe to use, is regulatory acceptable and the surface is suitable for accepting a printed symbol.

An important feature of these data carriers is error control (error detection and error correction). Linear bar codes are typically printed with a human readable form and if for any reason the bar code cannot be decoded, a human operator can key in the human readable data. These also incorporate a check digit, which allows any decode errors to be detected. If a decode error is identified or the bar code is physically damaged, a human operator can key in the human readable text. Clearly for 2D bar codes, this is insufficient, so robust error detection and correction schemes have been developed. These allow errors up to a certain proportion of the payload to be both identified and corrected, at the expense of requiring more error correction code words to be incorporated into the symbol. Composite code structures are also available that exploit a combination of linear and multi-row bar coding.

1.5.1 EAN.UCC bar code data carriers and application identifiers
A number of bar code data carrier standards (symbologies) have been adapted as EAN.UCC standards for carrying system numbers and identifiers. The capability of being able to use the numbering structures in data carriers that can also allow further data to be added and distinguished in a standardised way offers considerable flexibility in supporting item-attendant data handling and process improvement/innovation. The data carriers adopted for EAN.UCC system applications, that are now well established, presently comprise linear bar code symbols supported by the following standard symbology specifications (the rules that determine how a bar code is structured):

* EAN.UPC symbologies including UPC-A and UPC-E, EAN-13, EAN-8. These are symbologies specifically designed for omni-directional scanning at point-of-sale retail outlets and constitute the standard for use on items scanned in this way. The symbols may also be used on other trade items.
* Interleaved Two-of-Five (ITF) symbology for symbols carrying identification numbers on trade items not for scanning at retail outlets. The symbology is particularly suited for printing directly onto corrugated fibre-board and similar substrates. However, in contrast to EAN-UCC standard symbologies, the ITF is not exclusively licensed.

- UCC/EAN-128 symbology, a particular variant of Code 128, is exclusively licensed to EAN.UCC as the symbology supporting systems applications in which the system numbering and application identifiers are exploited. It is a variable length, alpha-numeric symbology offering considerable flexibility for identifying and handling item-attendant data. The symbols are not intended to be scanned at point of retail but within other areas of supply chain and industrial activity.

A reading device for EAN/UCC symbologies usually carries the facility to generate, on reading a bar code symbol, a symbology identifier to be transmitted along with the element string as a means of distinguishing between the different EAN.UCC data structures and those of other bar code symbologies. Such facilities are of course important for achieving automatic processing of data, particularly for transactions and EDI message formatting. The EAN.UCC system also defines a range of over 90, two, three and four digit application identifiers (AIs) and so provides a framework for supporting the identification of application measures (Fig. 1.7). AIs are also available to identify features such as logistics units expressed as a serial shipping container code (SSCC), batch and lot numbers, serial numbers, production and packaging dates to name but a few. A data format is specified for each AI to indicate the number and disposition of numeric and alpha characters. The AI to denote the identification of a logistic unit comprising the SSCC is 00, having the format n2 + n18, two digits (n) for the AI and 18 for the SSCC. An example of an alpha-numeric AI is for a lot number (AI = 10) having the format n2 + an..20, denoting two digits for the AI and up to 20 alpha-numeric (an) characters.

The AIs for measures are grouped into metric and non-metric trade item measures and metric and non-metric logistic item measures for parameters such as weights, lengths, areas and volumes. While numbering and identifiers

Extract from the list of EAN application identifiers (AIs)		
AI	**Encoded data content**	**Format**
00	Serial Shipping Container Code	n2 + n18
01	Global Trade Item Number (GTIN)	n2 + n14
10	Batch or Lot Number	n2 + an..20
13	Packing Date (YYMMDD)	n2 + n6
15	Minimum Durability Date (YYMMDD)	n2 + n6
21	Serial Number	n2 + an..20
30	Variable Count	n2 + n..8
310x	Net weight in kg	n4 + n6
400	Customer's Purchase Order Number	n3 + an..30
410	'Ship to – Deliver to' EAN.UCC GLN	n3 + n13
421	'Ship to – Deliver to' Postal Code with 3-digit ISO Country Code	n3 + n3 + an..9
...		

n – numeric number; an – alpha numeric

Fig. 1.7   Examples of application identifiers in the EAN.UCC system.

Example – RSS-14
stacked code

Example – 2D
composite code

Fig. 1.8   Examples of reduced space EAN.UCC data carriers.

constitute the means whereby traceability can be implemented, functional information sources with appropriate access codes are required to exploit the traceability expedient. Also required is a vehicle for partitioning information according to process and communication needs.

More recent additions to the EAN.UCC printable range of data carriers are the reduced space symbologies (RSS) and composite symbologies, examples of which are illustrated in Fig. 1.8. These symbologies have not yet achieved the impact that their potential offers but provide a useful extension to the printable data carriers available to support traceability. The RSS symbologies (which in effect comprise a group of RSS limited, stacked and expandable codes) are essentially high-density linear bar codes designed to accommodate the EAN.UCC system numbering, encoding up to 14 digits in a very small 'foot-print'. The expandable RSS symbols can accommodate additional data. The composite symbols comprise a linear (UPC/EAN, RSS or EAN128) symbol paired and in some cases data linked to a 2D symbol and printed immediately above the linear component. The 2D component is either a PDF417 or MicroPDF417 symbol, with data capacities ranging from 56 to 2,361 digits according to choice.

1.5.2   Radiofrequency identification (RFID)

Bar codes have two potential limitations relating to the line-of-sight constraint and the fixed nature of the data payload. Radio waves can, however, penetrate commonly used packaging materials such as paper, cardboard, plastic, and in conjunction with suitable semiconductor 'chips' can be configured to provide RFID data carriers or 'tags'. Such tags can be read from and written to using an appropriate radio-frequency carrier and modulation scheme and in the case of passive or battery-less tags, the RF carrier also powers the tag.

Although RF tags therefore appear to offer a number of advantages over bar codes, their use introduces other complexities and constraints that must be recognised. Firstly, RF tags containing semiconductor chips will always be more expensive than bar code technology, and it is difficult to see the business case for replacing bar codes on low-cost items. Secondly, the EAN.UCC system is a standard that is usable worldwide, with users in China for example being confident that they can scan an EAN bar code printed in the UK because the EAN.UCC standards define both the numbering system and the data carrier. RFID currently has five frequency bands that are used, but not all frequencies are usable worldwide and there are also regional differences in the allowable maximum power levels. ISO/IEC have standardised the air-interface for different tag frequencies under ISO/IEC 18000 parts 1–6 and both Data Protocol – Application Interface (ISO/IEC 15961) and Data Protocol – Encoding Rules (ISO/IEC 15962). Also being pursued with respect to the ISO/IEC 18000 series of standards is a part 6 development to accommodate electronic product code (EPC) air-interface protocol.

It is clear that RFID has great potential in many applications where line-of-sight or read–write data is required. For example, the animal tagging standards ISO 11784/5 define the code structures and technical concepts for the tagging of animals using low-frequency RFID tags. For many animals an ear-tag is used but for cattle, bolus tags, which lodge within the rumen, can also be used.

1.6   Linking item-attendant data and database information

Wireless communications and the Internet enable remote access to previously inconceivable quantities of information. Given our newfound freedom to communicate between remote data sources, the 'licence plate' approach to accessing information within a traceability system may be viewed as the single vehicle for accessing data held remotely. Every entity would be reduced to a unique number and all data relating to this entity would be held in a single database or a series of distributed and linked databases. All the required information would be accessible and updatable from any point at which the unique identifier can be entered to query the remote servers. Unfortunately this is not always a practical proposition.

When considering the appropriate use of connectivity in a given application there is a need to establish the balance between item-attendant data payloads and accessibility of data from remote databases through the available connectivity. It is often essential to guarantee a predictable response time to a remote request for information and to be able to safeguard the application against failure due to disruption or loss of connectivity. Where there is an opportunity or a requirement to exploit connectivity and access data remotely but performance constraints (such as a predictable and short transaction time) counter-indicate the use of direct information access, consideration should be given to the local caching of data or information in a suitable data carrier on the item itself. This will reduce significantly the demands on remote host bandwidth and reduce problems due to loss of connectivity or slow response times. By recognising the capabilities for using connect, cache and carry, independently or in combination with connectivity, considerable flexibility can be realised in distributed solutions for the management of item related data.

As with the data carriers for satisfying the vertical inter-linking needs to achieve traceability through use of a standardised approach to identification, so too there is a need for a standardised approach to data exchange and the accessing of information. A significant feature of any traceability system is the facility for communication and information exchange. Electronic data interchange (EDI) has for some time been applied as a fast and reliable means of achieving electronic, computer-to-computer exchange of information between trading partners with a supply chain legacy based upon the use of the EANCOM® language (a subsystem of the EDIFACT (Electronic Data Interchange for Administration, Commerce and Transport)). Some of the 47 message structures provided in this standard have relevance to traceability, including shipping notice (DESADV), product information (PRODAT), receiving notice (RECADV), transportation status (IFTSTA) and inventory status (INVRPT).

The advent of the Internet and mobile data communications have provided important new dimensions for communications and information exchange that can be readily applied for traceability purposes. They emerge as timely vehicles for helping to accommodate the developments in globalisation of trade and the growing diversity in supply chain structures, and developments are now in prospect for exploiting XML (eXtensible Markup Language) as a facility for supporting traceability communications.

1.7   The FOODTRACE project

FOODTRACE was a fifth framework European Concerted Action project, the aim of which was to establish a generic framework for harmonised supply, and cross-supply, chain traceability. The primary need[5] to be accommodated was a technology-independent, but technologically supportable, identification scheme for achieving traceability. This would allow developing countries the facility to specify traceability systems that allow migration from rudimentary supported systems to the technologically supported systems achievable in the more developed countries. The essence of such a scheme resides in agreement upon numbering and identification schemes, of which the EAN.UCC system is internationally recognised and, moreover, is supported by adopted technology for carrying data. However, it was recognised that the EAN.UCC legacy, although a core component of such a framework, would undoubtedly require further development to accommodate needs arising from developments in handling various aspects of system infrastructure development, including enhanced granulation of item identification and coding structures to allow identification of supply chains, primary identifier profiles, information sets and access control for multi-function traceability.

---

[5] Primary need – that need which if unfulfilled compromises the target solution even though all other needs may be taken into consideration within the framework.

1.7.1   Coding and access strategy

A significant work item within FOODTRACE was to define the overall information and access strategies for the generic framework. In defining a 'vertical' minimalist structure for using licence-plate coding links for and between supply chains the need was seen for coding that links the 'vertical' structure to the 'lateral' nodal structures for storing supply chain information relevant to the particular supply chain and the various traceability functions that need to be supported.

The data structures proposed for use within the supply chain traceability systems for inclusion in item-attendant data carriers comprise a set of item identification, information identifiers and access codes, including:

- **Supply chain coding** – codes to identify particular supply chains.
- **Nodal location coding** – codes to identify supply chain nodes and inter-nodal locations where significant processing of food items occurs. It is envisaged that these codes will use or be based upon the EAN.UCC numbering and identification system for location coding.
- **Item identification coding** – codes that will use or be based upon the EAN.UCC numbering and identification system. The need to extend this system is being considered to allow identification of individual food items, from ingredients to products and product variants. Each component in the realisation of products will need to be identified and linked to the parent source or sources of constituent parts and any immediately previous nodal coding.
- **Information set coding** – codes to distinguish nodal information sets corresponding to particular traceability and support functions. It is envisaged that these codes will feature as possible extensions to the EAN.UCC numbering and identification system application identifiers.
- **Access coding** – codes to allow access to information sets, with a default level to general information within sets and priority levels to control access to particular components of information.
- **National location registers** – to register at national level supply chain and nodal location codes, together with traceability function codes. It is envisaged that such registers would be accessible through the Internet and eventually offer international linkage for traceability purposes.
- **Label-based product identifiers** – machine-readable coding for information access through the national location registers, allowing access to general information. It is envisaged that the label coding would include the item identification, supply chain code and the last nodal location code. By reading the code the user would be linked to the last node through which the product came and to the general information relevant to that item. The register site would support access to other information in the supply chains through use of appropriate access and traceability function codes. This would allow traceability investigators with the appropriate access codes to access particular traceability information stored in nodal-based information sets along the supply chain. Similarly, commercial information

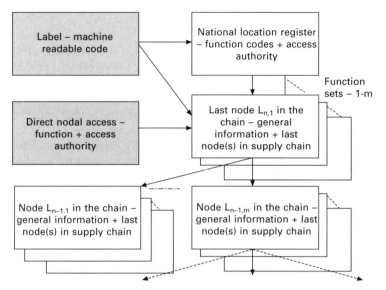

Fig. 1.9   The structure of the FOODTRACE system.

may be accessed by means of priority access codes. Item information would be accessible at any node within the supply chain using item and access codes to constituent information sets.

By way of example, a label upon an item taken at the point of sale or consumption would allow access to a National Location Register or last node at which the item was realised, wherein general information on the item could be obtained, together with last nodal location at which the item was realised if not included in the label. Access to the nodal location would provide information on what went into the product or how it was processed. It would also identify the nodal locations corresponding to each of the component parts of the item or the processes linked to this stage in the supply chain. Access to other nodes within the chain would similarly provide constituent and process information corresponding to the item or items concerned. The level of available information would be determined by the access codes available to the investigator and the function information sets supported. Rapid access tooling would allow rapid listing of constituents and locations corresponding to products produced and transferred through the chain (Fig. 1.9).

1.7.2   The need for a universal data appliance protocol
In the absence of a universal data carrier or a range of standards-supported data carriers there is a need for a universal protocol and interface platform to accommodate the diversity of data carrier and other item attendant technologies available for realising traceability systems. The disparate nature of the

technologies and the associated products present problems in interfacing different products to different software systems. A common protocol to accommodate these differences is therefore seen as an essential requirement for optimising supply chain systems. Moreover, the need is seen for a flexible protocol that can accommodate the consequences of change and facilitate migration, as appropriate, to more advanced systems and systems intelligence.

The need for a universal method of integrating or connecting different item-attendant or associated appliances into systems is already being seen and accommodated through proprietary developments. From a generic traceability standpoint it is important to consider the need for an industry-wide standard. The basis for this assertion resides in the need to specify:

- a common interface between back-end enterprise software and a disparate range of item-attendant data collection devices
- forward compatibility for new data collection devices and support technologies
- plug and play capabilities with respect to bar code, RFID and other automatic identification and data capture devices
- remote network management with optimised network reliability.

Developments in the XML-based approach to messaging may be seen as a highly extensible solution in this consideration, supporting the transfer of multiple elements in a single message.

Other features include:

- simple command structure that allows easy configuration and administration
- automatic recognition and registration of new devices on the network supporting a unique level of interoperability with different data collection devices
- flexible and extensible device facility that enables all the relevant characteristics of the devices on the network to be automatically described, including properties, methods and events
- automated monitoring of device availability and status
- alarm events to indicate problems with device functionality.

Herein lies the facility for a fully integrated traceability infrastructure.

1.8   Conclusions

The FOODTRACE generic framework, based upon the considerations outlined earlier, is essentially an attempt to provide an holistic approach to traceability of open systems by providing a significant degree of harmonisation and inter-operability but without sacrificing privacy of information and confidentiality where such is required.

Achieving a generic framework for harmonised traceability within and across supply chains is an important requirement. The approach being adopted through FOODTRACE distinguishes the minimalist vertical and lateral infrastructure and strategic, standardised coding of items, supply chain nodes, information sets and access systems. A significant legacy is in place to assist this process, including standards for numbering and identification and electronic data exchange. By building upon this legacy the opportunity is seen to realise the aim of the FOODTRACE initiative. Moreover, by appropriate attention to legacy and migration strategy the framework can be developed to accommodate existing traceability systems and future requirements with respect to traceability functions and emerging legislation through information set coding and access structures. Through developing and applying appropriate process methodology the traceability infrastructure may be developed in a way that adds value to supply chain processes with attendant benefits with respect to food production, distribution and the reduction of waste.

The follow-through deliverables of FOODTRACE include a set of guidelines for assisting supply chain developers in structuring traceability systems to fulfil their individual requirements. Being generic the framework offers a scheme for developing solutions that are supply chain specific yet offer interoperability as and where required, together with scope for process and supply chain innovation.

2

Using traceability systems to optimise business performance

F. Verdenius, Wageningen University and Research Centre, The Netherlands

2.1 Introduction: the FoodPrint approach

Traceability in food chains is often seen as a means to ensure food safety, to minimise the impact of food incidents, and to manage liability issues. Viewed this way, traceability is often seen as an imposition requiring investment but not contributing to profits or competitiveness. This is one of the reasons why traceability has been introduced slowly in the food sector. FoodPrint is a systematic approach for analysing and designing traceability systems that takes the business goals of the food companies as a starting point. In a strategic traceability analysis, the goals of the system under study (company, production chain, or even an entire sector) are carefully studied and connected to potential benefits of traceability. Subsequently, the current traceability information system is modelled and traceability bottlenecks in relation to the business goals are surveyed. This forms the basis for the design of an improved traceability system.

In this chapter, FoodPrint is considered as a means for optimising the business performance of a food company or food chain. Firstly, some basic concepts of traceability are introduced. Then some prominent trends that lead to increasing traceability requirements for the food sector are described. Societal and legal developments, technological and economic evolutions in the food sector and the rapid progress of traceability technology together determine the current attention for traceability in the food sector. In section 2.5, the major FoodPrint concepts, the phasing and a number of the tools that support the system analysis and development, are described. Section 2.7 describes some of the experiences that have been gained in applying the FoodPrint approach, after which there are some concluding remarks.

2.2   Key concepts in traceability

Traceability is the ability of a company, a chain or a sector to trace the history of a product through a production and distribution chain from its initiation until its consumption or application. A product entity in this respect is an identifiable individual product: not milk in general, but the individual pack of milk that is standing in front of me at this very moment. The history gives the sequence of production and distribution phases that a product has gone through. What types of phase are of interest, however, depends on the purpose of operating the traceability system. As explained in this chapter, there may be a number of different reasons for setting up a traceability system. The initiation of a product entity is the moment that it is created as an identifiable product unit. In the case of the pack of milk, initiation starts at the moment the milk is tapped from a large tank into the carton. A product entity comes to its end when the entity ceases to exist, due to consumption or application in another product. In the case of the pack of milk, this can be the consumption of the milk or its application in a dough for baking a pastry.

2.2.1   History of traceability

Traceability comes from two different directions: logistics and safety management. In logistics, increasing efficiency of industrial methods has urged the need for improved efficiency of production and distribution logistics. This trend is omnipresent in almost all industrial sectors, but it certainly has been accelerated in classical industrial sectors with a high pressure on cost reduction such as the automotive and aerospace branch and in the specialised sector of couriers. Since WWII, more attention has been paid to logistics with the main goals being the reduction of financial losses due to stock keeping, planning slack and process deficiencies, and an increase in flexibility to offer responsiveness to consumer demands that is typical for mass industry. Resulting supply chain strategies, such as just-in-time (JIT) are practically impossible without reliable traceability systems, combined with sensitive early warning systems. The standards for traceability that have been developed in the automotive and aerospace industry have by now been adopted by many industrial sectors.

A second driver for traceability that emerged simultaneously in the various technological sectors (automotive, aeroplane and aerospace industries) and in pharmacy, is the concern for product safety. Years of experience with safety incidents, liability issues and public health risks have resulted in advanced traceability systems in these sectors. Such systems ensure the ability for a focused recall (based on serial numbers) in case of a critical product error. Moreover, many of these systems offer the ability to analyse and diagnose occurring safety problems. When, for instance, in 1992 an El-Al Boeing 747 crashed in a residential area in Amsterdam, the detailed data of the aeroplane could fairly rapidly be drawn from existing records.

Information on the type of engine, the type of blades in the turbofan, the types of bolts used to secure the engine to the pylons, defects during recent security and maintenance checks, and details on the operation history was available almost instantly. As a result, basic technical questions could be answered rapidly. In pharmacy, concern for failures in production, handling, or storage has led to a detailed traceability system. It is a well established practice that packages are given lot numbers that may be used to identify logistic, storage condition and manufacturing data relating to the individual products.

From WWII until the end of the 1980s the main focus in agriculture and food supply has been on large-scale low-cost production. Available technologies for making products traceable, as developed in pharmaceutical and technological industries did not percolate massively into food supply chains, although large producers, especially of A-brands, did use traceability schemes as tools for quality management. During the nineties, the understanding emerged that food chains are very vulnerable to malfunctions. The cry for effective measures against food incidents has increased. Apart from many 'annoying' incidents having little impact, some major plagues have struck the food sector, e.g. BSE, foot-and-mouth disease and a number of dioxin poisonings.

2.2.2   Technology

Traceability presumes the ability to identify products, to register product presence and, when required, product attributes and process characteristics, and to document and analyse the registered information. For these traceability functions, solutions have to be implemented by means of specific technical systems. In order to select the appropriate systems, it is necessary to consider the exact goals that are pursued in the traceability system. Traceability systems exist for three basic orientations: location, conditions and quality (see Table 2.1).

Table 2.1   A framework for traceability technology

Function	Orientation location	Condition	Quality
Identification	Alphanumeric tag, bar code, passive RFID	Data logger, active RFID, process database	Metabolomics, genomics, biochemistry
Registration	OCR reader, bar-code scanner, RFID reader	RFID reader, data logger reader, TTI visual inspection	Automatic or manual measurement, micro-array assessment
Data processing	ERP, MES, dedicated traceability system quality system	Chain monitor system, quality development models, process quality system	Product quality system

In Table 2.1, three orientations are listed: location, condition and quality. The location orientation focuses on logistic aspects of product flows. The main issues here are time, place and logistic phase of a product: where was the product at a certain time and with what logistic status? Product flows, either individual products (e.g. cattle) or product lots (e.g. a silo content), are equipped with an identity and that identity is registered at specific moments in the process. Diversion and conversion of product flows are reasons for issuing new identity labels.

The condition orientation corresponds with process aspects of product flows, i.e. what conditions and process settings have been applied during the production and distribution of products. Process systems document the process conditions and process settings in a production process. Data loggers can monitor the conditions products experience during logistic phases. The collected information is used to assess the extent to which chain actors have fulfilled optimal process conditions. Typically, the definition of relevance of conditions depends on product characteristics. For the production of microchips, the amount of 'contaminating particles' in the process environment constitutes a critical condition. For fresh fruits, the concentration of ethylene and the ambient temperature are critical factors, whereas, during the production of convenience foods, the microbiological conditions may be critical.

The quality orientation relates to the product quality aspect of product flows. Here, the actual product quality is relevant, under the assumption that the quality is dynamic and depends on the actual chain performance. This is typically the case for fresh and perishable products. Traceability systems with a quality orientation are typically focusing on the development of quality in the chain. The quality aspect needs to be defined and a quality measurement technology needs to be selected. To fully utilise this option, two conditions have to be fulfilled:

- Definition of quality. As indicated above in the definition of relevant conditions, the definition of relevant quality aspect will strongly depend on product properties and customer expectation.
- Measurement of quality. Many quality aspects that are important from a customer point of view (taste, keepability, flowering stage) cannot yet be measured directly. Modern measurement technology can help to assess quality aspects more directly.

The vertical axis of Table 2.1 enumerates the three main functions of traceability systems: identification, registration, and data processing. Identification of a batch is the coupling of the physical product or lot to the information that belongs to that product or lot (in most food applications, lot identification suffices). The identity can be physically attached to a product by means of a textual code, a visual code (e.g. a bar code) or an electronic code. These codes are carried on a label, an RFID tag or any other type of tag. Such a code substantiates the identity of the product (batch). The tag substantiates the product identity. Elsewhere (FoodTrace, 2005), this is referred to as the primary identity (expressed in the product itself) and the secondary identity (expressed in the product tag).

Although, in principle, a pen-and-paper system can do the job, in many cases chain actors prefer to use (semi-)automated equipment to support traceability. In that case, they have to select a coherent set of tags, readers and data processing software. Elsewhere in this book (Chapter 10), a vision is presented on tracking and tracing technology (Vernède and Wienk, 2006). The discussion of technology in this chapter is limited to a functional view. One important aspect is that a feasible tagging schema is able to recognise levels of product units. In the EAN coding schema, for instance, unit products, packed products (e.g. boxes, bags; code type), pallets (code type) and containers (code type) are distinguished.

Registration of a product entity allows the scanning of the products' secondary identity. By reading or scanning the code on the tag, the product identity is accessed and can subsequently be used to access product (lot) data in a database. Registration information comprises a minimum of product identity, registration time, registering resource and process phase.

The content of the data processing partially depends on the goals that are to be achieved in using a traceability system. The initial processing, storing the registered data in the database, is standard for all types of systems. In the functionality of additional processing modules, the specific goals of the traceability are reflected. In existing food chains, the most often encountered goal of a traceability system is to control the impact of food safety incidents. This implies a focus on assessing the potential scope of an incident and on performing a recall. The fear to be seen as responsible for a food safety incident and the increasing pressure of governments to control food safety by enforcing traceability systems (e.g. the European general food law, EC 178/2002), triggers the interest in these kind of applications.

Traceability systems are, however, not limited to food safety and liability issues. In logistics, reducing the vulnerability of a chain for external influences (e.g. traffic problems, production delays) can be a goal for a traceability system. This is, amongst others, often found in couriers and in just-in-time chains (e.g. the automotive and aerospace industries). In quality management of food production systems, traceability systems can help to understand quality development of products. In this case the condition or quality orientation of the traceability system becomes of vital importance.

2.3   Traceability in food chains

The current interest in traceability has been inspired by the impact of various recent food safety incidents, fraudulent origin indications (both Europe) and the fear for bioterrorism (US) after the 9-11 turbulence. This has led to enhanced regulations for traceability. In Europe, these regulations are, amongst others, defined in the General Food Law (GFL) (EU 2002), that enforces a one-up-one-down regime as the minimum traceability requirement. In the United States, the Bioterrorism Act of 2002 (http://www.fda.gov/oc/bioterrorism/bioact.html) requires, amongst other things, the registration of food handling, producing and storing facilities that directly supply US destinations and country of origin labelling of food products as basic traceability measures to limit the risk of food adulteration.

2.3.1   Defensive versus offensive

In a recent study in the Netherlands (Imtech, 2004), it was concluded that most food companies postpone innovations as long as legally possible. Consequently, innovations are only accomplished when an obligation enforces companies to introduce new technologies or new operating procedures. In addition to legal obligations, the food sector is characterised by a lack of balance of power. Chain partners close to the consumer, for instance producers of branded products or retailers, impose stringent requirements onto the (upstream) chain. This has resulted in traceability requirements for producers that anticipate, and are often more stringent than, the applicable legal obligations. In line with the mentioned attitude towards innovation, companies deploy a reactive, or defensive, strategy towards the introduction of traceability, in the sense that they wait to introduce traceability until they are forced to do so. When introduction has become unavoidable, organisations will invest the minimum that is required.

In contrast with the externally motivated defensive introduction of traceability is the proactive, or offensive, approach. Here, traceability is employed as a means to achieve particular business goals. The starting point is the recognition in an organisation that a specific goal needs to be achieved and the awareness that traceability can play a role in reaching this goal. In many cases, organisations view traceability as a technical solution to an externally imposed problem. The change of attitude that is needed to accept that a traceability system can be beneficial to a company is not easy to make.

2.3.2   Risks versus challenges

Traceability is often associated with risk control and crises management, especially in the area of food safety risks. Being able to pinpoint a food hazard to a specific lot and to link this lot to suppliers, clients, processes and raw material lots clearly helps to manage incidents.

2.3.3   Traceability for risk control and crisis management

The major motivation for companies in food chains to introduce traceability is the concern for food safety. By labelling product batches and by registering destination and origin information at batch level, potentially hazardous infected or contaminated products can be easily removed from trade and consumers can be called to return their product. Moreover, the introduction of traceability makes food suppliers, food producers, processors and carriers accountable for their responsibility for the incident that causes the need for corrective actions. In this way, one could argue that traceability also has a preventive effect.

2.3.4 Quality-oriented traceability

A next step in traceability is set when the information that is gathered in the chain is not only used for post hoc consultation (as in the case of a recall action) but also for active management of the process. This requires an extension of the types of information that are recorded. Potentially (and this development is already occurring in practice), the tags are upgraded from passive carriers of identity information towards active devices that document features of the process, such as temperature, relative humidity and, maybe in the near future, gas concentrations, vibrations and mechanical shocks, or microbial contamination.

2.3.5 Value traceability

Once it was reckoned that the information in a traceability system could be used to accomplish particular company goals. Instead of serving goals of external actors, new directions for traceability could be explored. Based on information on the history of products, processes can be made more efficient, leading either to a better product quality or to improved process settings. With this in mind, companies can broaden their view to adding value to their business by means of traceability systems. This broader view on traceability has been indicated as value traceability (Pape *et al.*, 2002).

2.4   Factors affecting traceability systems

Traceability, the documentation of product flows, is a means to make the history of a product available for inspection. To achieve this, the flow of a product in a production and distribution chain is registered and stored in a database. The recorded information can be used for various business functions. Primary traceability functions include real-time access to the current product location, product status, the responsible actor or employee for the product, the destination of a specific product and a list with the batch numbers and suppliers of the raw materials used to produce a certain product. Secondary traceability functions include recall, generation of product genealogy, generation of production data, environmental conditions and product quality.

Traceability is not a self-standing goal. No one business wants traceability without having specific objectives in mind. Historically, the needs for safety, minimisation of liability and optimisation of logistics have been the motivators for developing and refining traceability concepts. Main motivators to implement traceability in food chains are protection against the impact of food safety incidents and traceability as license to deliver.

It is important to consider that the traceability of food products is just one development in the food sector. The sector is confronted with many other challenges. From the viewpoint of traceability the relevant areas where developments take place are technology, food chains and society.

2.4.1   Societal trends

Societal trends are complex, compound and open to various interpretations but it is possible to distinguish consumer trends, civil trends and legal and governmental trends. Modern consumers here become more demanding in such areas as luxury, convenience and low tolerance towards quality deviations. Modern production and distribution chains rely on a large transport capacity, both for obtaining ingredients over long distances, and to allow low-cost processing to take place in other countries. As a consequence, modern consumers are often ignorant of the origin and history of their food products, the source and calibration of ingredients and the processing history of the product. Furthermore, convenience, ready-to-eat and minimally processed products, with higher food safety risks, increasingly penetrate the market. At the same time, consumers demand an acceptable price-quality ratio (Sloof *et al.*, 1996). Due to these developments, consumers increasingly lack detailed knowledge on the products they buy and cook. How consumers balance price and quality is still subject to debate. Some surveys claim that consumers prefer quality to product price, whereas other results seem to indicate the opposite.

Simultaneously, the social and individual tolerance against deviating product quality decreases. Consumers (and retailers on behalf of consumers) expect to encounter the highest food safety levels, even when they themselves treat their products sub-optimally. When offered the choice, consumers avoid risks in their purchase behaviour but, simultaneously, often do not show an explicit interest in product history or traceability.

Individuals do not act only as consumers but also as citizens with social and ethical concerns. People feel uncomfortable about some production methods. As an example, GMO products can count on little sympathy, at least in Europe, where civilian concern seems to be stronger than the consumer interest and where civilians sympathise with organic, fair-trade and animal friendly production methods.

Governments on the one hand have to relieve the burden on consumers and, on the other hand, they have to guard the economic interests of companies. In the area of traceability, this leads to a regulation that is not always implemented unequivocally. In the European beef sector, labelling obligation was introduced in 2002, requiring the announcement of specific information on the origin of the meat. The introduction of the labelling was motivated by the social unrest caused by the BSE incidents in the 1990s. In spite of this legislation, inspections in later years showed that a substantial proportion of beef products did not comply with the labelling requirements.

2.4.2   Trends in food chains

In the food sector, a number of developments can be observed. First, the complexity of food chains has increased substantially over the last few decades, leading to an increasing vulnerability as well. Raw materials, for feed and food producing industries, are grown world wide and transported to processing plants. Product lots undergo multiple processing stages and secondary product flows are increasingly used as raw material for other food or feed production. The impact of these factors becomes clear when incidents occurred. In a recent incident in the Netherlands, milk was contaminated with dioxin. The cause was the use of marl clay in a sorting process for potatoes for the French fries industry. The peelings are used as a feed component. As a consequence, dioxin, in non-harmful doses, has been identified in many products. In spite of the increasing quality standards, safety measures and traceability requirements, this kind of incident keeps appearing. Apparently, the overview of a production chain is limited and the number of unknown factors is too large to allow all possible risks to be taken into account. Therefore, due to increased detection thresholds and reduced tolerance, the number of detectable incidents increases.

A second development is that products are innovated. The trend to convenience/ready-to-eat has been mentioned. This kind of product is more susceptible to, e.g., microbiological contamination. The ability to monitor and assess the historical storage conditions may help to monitor quality development, but only when the monitoring covers all relevant aspects with sufficient detail.

2.4.3   Technological trends

Technologies for traceability show a rapid evolution. This is caused by the pressure to come up with new solutions due to food crises and by the rate of innovation in technologies such as new materials and new technologies, leading to miniaturisation, enhanced functionality and integration of traceability with other technologies, e.g. ICT and quality care. Recent developments include: miniaturisation, (inter)activity of tags, e.g. through RFID and vision systems, integration of sensor technology with tags, DNA and genomics technology, and the inception of technological standardisation for several of the techniques (e.g. RFID).

These developments have effects in three major areas. First, the omnipresence of traceability, the increasing information that is included in traceability systems and the expanding computerisation leads to global connectivity of chains. At this moment, it is already possible to obtain information on product origin and it will not be long before details on handling, growing conditions, use of pesticide and herbicides etc. will be available on line. This will enable industry, trade and consumer to be more aware of quality and risks.

A second trend is the ambivalent movement towards both centralisation and decentralisation of traceability information. On the one hand, there are systems that combine the traceability information in a central database, for various chain phases and sometimes chain partners. An example of such a system is the Chainfood/GroeiNet approach (www.chainfood.com, see section 2.7). In this system, chain partners provide their information, or references to their information, through an internet interface in a central backbone system. Common functions, including traceability, chain optimisation and data analysis, are implemented to operate on this database. On the other hand, there is a trend to offer local traceability. In this case, there is no central data available and traceability is realised in the function of collecting and combining the relevant product information. In such a setting, a local database can be accessed through a communication infrastructure.

The final trend, with probably the most impact on the food sector, is the integration of traceability in other business functions. Originally, traceability has been motivated by the desire to control the impact of food safety incidents. Both governments and industries now start to perceive traceability as a tool that can be used for many purposes, including food safety management, food quality management, logistics, process efficiency, cost control, marketing and many others. Consequently, traceability can be implemented as a separate business function, but it can also be part of ERP, logistic, commercial or quality management solutions.

2.5   The FoodPrint model for developing traceability systems

With value traceability, traceability becomes a tool to realise business goals. This requires an approach to designing traceability systems that takes these business goals as the starting point. FoodPrint is such a systematic approach for designing tracking and tracing systems. It integrates experiences from quality-related tracking and tracing with value traceability projects in food chains. It takes the business goals of an organisation as starting point and uses these as a guiding principle when translating legal and commercial traceability requirements into practical solutions. Traceability thus becomes an instrument to realise the business strategy and allows companies to reduce product and process losses, to realise commercial advantages and to develop new products.

In the discussion of the FoodPrint approach, the components of a traceability system are described. Next, the motives that companies may have for

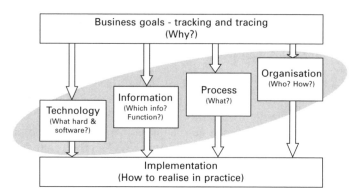

Fig. 2.1   Components of traceability systems and the main questions in each component. Before design of the four components lies the analysis of business goals; after it lies implementation.

implementing traceability systems are explored. Then a typical development process is outlined, concluding with the discussion of a number of development tools and concepts.

Implementing traceability is sometimes perceived as the introduction of tagging and registration technology in an existing production or distribution process. In fact, the introduction of technology is the final step in the introduction of traceability, which is preceded by defining and designing traceability in the organisation, the production or distribution processes and the information system. Traceability systems can be perceived as having four components: organisation, process, information and technique. Figure 2.1 shows these components in the overall development process of traceability systems.

2.5.1   Organisation
The organisation level is an essential aspect of traceability, even more so when the targeted traceability exceeds the borders of individual companies. Although current legal requirements on traceability only require traceability at the company level, modern complex food chains and networks already go beyond company borders. Large retail organisations, multinational food producers and international integrated food chains (both in integrated businesses and in collaborating companies in a production and distribution chain) have implemented traceability systems that pass company borders. At the organisation level, arrangements are found that are necessary for obtaining traceability. Responsibilities for product and information handling, agreement on (the synchronisation of) procedures, definition of traceability information and the definition of information exchange standards in a chain, decision making and traceability goals are established at this level. Traceability goals can be focused on individual companies as well as on entire chains or networks.

Aspects at the organisational level that need to be specified include:

- **Scope**: description of the part of the chain that is covered by the traceability system
- **Allocating responsibilities**: who is responsible for products, how is responsibility transferred to the next link, how is liability transferred through the chain?
- **Allocating cost and benefits**: costs and benefits of traceability are not equally spread over the chain. Extrapolating earlier research, it can be expected that the balance is more positive at the end of the chain. Chains that agree on a shared traceability effort need to negotiate how costs and benefits can be more equally shared over the chain. Sharing costs, information, or benefits can help to do this.

2.5.2   Process

At the process level, the production process is defined by describing the various steps in the process, including product flows along the chain. Per process step, information requirements and information flows are defined. Moreover, control processes need to be documented.

Although traceability is primarily a matter of information flows, the process organisation in detail can influence traceability. Traditionally, processes in food chains have been designed with the aim of efficiency and cost reduction and not on traceability. As a consequence, many processes have inherent problems with detailed traceability. Examples are the mixing of raw materials in bulk silos, the uncontrolled mixing of product flows in larger logistic units and the lack of product registration at points of converging and diverging product flows. In many cases, when more stringent traceability limits are required, the process layout needs to be adapted. In Chapter 5, the process and bottleneck analysis are described in more detail (Koenderink and Hulzebos, 2006).

2.5.3   Information

The traceability of product flows fully depends on the ability to process information. Based on the process model, a specific traceability information model can be constructed. The information model specifies the structure of the information to be collected. This model can be based on a sector specific or generic reference model, when available. The model defines the information entities that are required for realising the traceability goals that have been defined for the company. Together with the process model, the information model constitutes the traceability model (Koenderink and Hulzebos, 2006).

2.5.4 Technology

Traceability can deploy a large variety of technologies. In many discussions on traceability, technology is even the primary point of discussion. In the FoodPrint view, the technological choices form the final step of the design and implementation process. Technology for traceability, especially in relation with product quality, is discussed in detail in Chapter 10 (Vernède and Wienk, 2006).

2.5.5 Motives for traceability

What are motives for the introduction of traceability concepts in food industries (Fig. 2.2)? In a recent survey (Imtech, 2004), it was noticed that many actors, and not only the small and medium companies, are triggered in investments mainly when government regulations force them to do so. For traceability, it is the same. Experiences over the past few years with regulations in the EU on traceability, for instance on the traceability of beef products (*Het Financiële Dagblad*, 2002), show that the implementation of traceability concepts is slow, even when the availability of traceability facilities is mandatory. Apparently, the perception of traceability by industrialists is one of investments with moderate benefits for themselves.

How do motives relate to applications (Fig. 2.3)? In terms of motives, the perception focuses on external or defensive motives. When realising the objectives that result from these motives, there is no clear benefit to be obtained for the investing company. By making the investment, the company defends its position in the market. The investments serve a license-to-produce, or a license-to-deliver, but they do not lead to a clear return in terms of an increased turnover, an increased margin or a better position in the market. Opposite to these defensive motives there may be offensive motives that help to improve the position of the company. These offensive motives concentrate on cost reduction and return maximisation.

Cost reducing motives include the reduction of product cost by a better

External motives	Cost reduction	Return maximisation
Customer demand	Product cost reduction	Increase of market volume
Chain partner requirements		Broadening of the market
Government regulations	Process cost reduction	Added value

Fig. 2.2 Grouping of motives for traceability.

Fig. 2.3   Motives for traceability in applications.

conversion of raw materials into products, product loss reduction and improved control of quality development of raw produce. Process cost reduction may include reduction of resource cost (e.g. energy). Return maximisation can be realised by adding value to products, for instance by being able to substantiate product claims such as organic, regional origin, allergen free and child labour free. Return maximisation also includes broadening of the market, for instance by positioning specific variants of the same product (taste variants, regional versus cheap variants). Finally, increasing market volume is an efficient way of maximising returns. This can be realised by using traceability information to pinpoint batches to the right location.

Consequently, FoodPrint starts with a thorough analysis of the business goals of the food company or food chain of interest. These goals are linked to concrete and measurable tracking and tracing targets. Next, the current situation is analysed and, by means of a bottleneck analysis, the future situation is evaluated. This finally results in system design which can then be made operational by an ICT system integrator.

During the process, FoodPrint differentiates the following modules: process, organisation, information and technology (see also Fig. 2.1).

2.5.6   Traceability systems

A traceability system is a coherent set of concepts, tools, working procedures and equipment that enables the tracking and/or tracing of goods in a production and distribution environment. Typically, a traceability system supports a minimal set of functions. Three functions are required to establish, record and store product and process data:

Product identification
Product identification is realised by using a unique product characteristic (e.g. DNA, spore-elements, antibody markers) or a product number, attached to the product on an external tag (e.g. bar-code, see http://www.ean-int.org/index.php?http://www.gs1.org/barcodes.html), 2D-dot matrix code, radio frequency identifier, see http://www.aimglobal.org/technologies/rfid/). The former is sometimes referred to as the primary identifier and the latter as secondary identifier (www.euFoodTrace.org). When collecting individual products in logistic units (e.g. box/pallet/bin and container), a hierarchy of identification systems can be used, as can be observed in the EAN system for coding products (EAN: GTIN), pallets and containers (EAN: SSCC)

Product registration
At relevant positions in the process, identification codes of products have to be registered. Relevant points are, at least, the points in the process where product flows converge and diverge. This allows the coupling between process parameters and product identity, as necessary to reconstruct product quality implications as encountered later in the distribution chain. The registered product code, the time stamp and the identification code of the location/process phase identification, provisionally augmented with additional information on conditions, product state and product characteristics, are recorded in a database.

Information processing
The product registration information is stored and, when the storage of information is in a database system, analysis and retrieval of relevant information is maximally supported.

In addition to these three basic functions, the analysis and retrieval function can be developed, related to the legal traceability requirements and specific business motives for traceability that have been identified for the company:

- ***Localisation of product***: often indicated as the tracking function. Provides an indication of the physical location, process status (waiting, processing, in packaging station, finished, delivered), under whose custody a product resides, and in which process stage (if required). This function fits to logistic tasks, such as (re)planning of transportation, cross docking and warehouse management.
- ***Reconstruction of the product genealogy***: A product genealogy is the record of converging and diverging product flows that indicate on the

production level how a product has been produced. It enables the trail to be followed from raw materials and intermediate products, through the various production stages, showing where product (batches) have been combined and divided. This is all coupled to product batch identities.

- *Reconstruction of product quality*: When a traceability system includes the registration of product (quality) state and process conditions, it is possible to assess the development of product quality. Getting a better impression of product quality may help to improve process planning, allocation of resources to batches and allocation of product batches to customer destinations.
- *Product recall*: In the case of a food safety incident or a food quality incident it may be necessary to withdraw a product from the market: a product recall. The detection of the current whereabouts of potentially hazardous products, eventually followed by automatic or manual actions, relies on the identifiability of products and the documentation of products flows.

On top of an infrastructure that allows these activities to be performed, additional functions, such as condition monitoring, quality reconstruction, quality prediction and various logistic, planning and administrative extensions can be realised.

2.6   Phases in the development of a traceability system

In the previous sections, it was seen that, for the introduction of traceability, it is important to formulate the specific motives for traceability for a chain or company. These motives form the basis for the operational requirements that have to be fulfilled. During the development process, the system to be developed is compared with the requirements. The results of traceability research over time have resulted in a number of tools and concepts for the development of traceability systems. These are collected in a systematic stepwise development approach for traceability systems: FoodPrint. FoodPrint distinguishes the following phases in the development of traceability systems.

2.6.1   Strategic traceability analysis

In Europe and the USA, with the implementation of the general food law (EU 2002/178) and the Bioterrorism Act of 2002, respectively, traceability of food products has become a requirement for all actors in the food sector. Implementation of traceability to comply with the imposed regulations is from a business point of view, however, not an investment that is commercially justified as it is an investment that has no benefits and it leads to an increase in costs. These costs are both fixed, in terms of additional equipment to be installed, and variable, in the form of tagging costs, the costs of operating

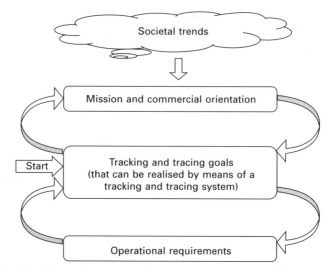

Fig. 2.4   Schematic representation of the strategic traceability analysis.

traceability systems and the costs of implementing additional procedures in the production chain. Customers, starting with consumers and retail, however, are not willing to stand an increased price. Consequently, the introduction of traceability seems to nibble at the margin of companies.

In order to get a clear and complete picture of the motives for traceability, the strategic traceability analysis is performed. In a three stage process (see Fig. 2.4), the operational requirements of a company, with relation to a traceability project, are derived from the mission that a company has formulated, or the commercial orientation that a company has developed. It is very hard to see the direct relationship between company mission and operational requirements for traceability. To facilitate the reasoning project, the discussion is best started by focusing on the concrete and actual reason for starting the project. In Europe, the mandatory requirements as imposed by the GFL form such a reason. Formulating these goals may lead to a first indication of the mission/commercial orientation that companies deploy. The level of tracking and tracing goals in general is very down-to-earth.

At the level of mission, companies can be classified in a limited number of company orientations:

- Consumer: Companies that are basically motivated by the desire of consumers. Typically, these are the companies that are strongly connected to A-brands.
- Chain/customer: Companies that are basically motivated by the desires of chain partners (often: retail) and direct customers.
- Internally: These companies are mainly driven by the chances to optimise their own operations.
- Supply chain: Companies that are focused on managing their upstream processes.

These orientations are adjacent to some of the integration levels of the Supply Chain 2000 (SC2000) framework of the CLM (Bowersox *et al.*, 1999). We slightly modified the class definitions and their content to fit the traceability needs of actual companies. As in the SC2000 framework, the orientations are not mutually exclusive: a company may exhibit characteristics of several classes. The membership of a class has direct consequences for the offensive and defensive motives that companies have for implementing traceability. A producer of A-brands, with primarily a consumer orientation, will be focused on traceability motives that give added value. A primary supplier, on the other hand, with a primary orientation on complying with chain partner requirements, will be triggered by the chain partner requirements, probably combined with cost reduction motives.

How to get insight into the motives for traceability in an actual chain or company? Depending on the complexity of the target organisation, an analysis process is accommodated. In complex chains, for instance for the horticultural sector in a specific country, such an analysis is a project in itself. Over the years, two special forms of strategic traceability analysis have been developed: the FoodPrint Quick Scan (FQS) and the SME scan FoodPrint Direct (FPD).

The FoodPrint Quick Scan, FQS, is typically suited for one single company or an uncomplicated food chain. In a short, condensed, process, the current situation, bottlenecks and traceability goals are described. In the process, it is the goal to get a shared perception of the traceability goals to be reached. This includes both the goals of individual stakeholders and the common goals that are shared by many or all actors in the chain. Typically, the throughput time of such a process is two to six weeks. In this period, two workshops with all chain actors play a crucial role. The first workshop focuses on describing the current situation and the perceived traceability bottlenecks. The second workshop defines the individual and shared traceability goals.

FQS typically aims at complex companies and singular food chains. For many small and medium enterprises (SMEs), such a process is oversized and, consequently, too expensive. To meet the specific needs of SMEs in complying to the European legislation (especially the General Food Law (GFL, EU 2002/178) FoodPrint Direct offers SMEs the possibility of obtaining a quick and shallow analysis, with the emphasis on three issues:

1. Does the enterprise currently comply with the GFL requirements?
2. If not, what improvements are needed to comply with the GFL requirements?
3. What offensive motives are necessary to profit from the introduction of traceability systems?

The results of this on-line analysis, which takes a maximum of 30 minutes to complete, can be used as the starting point for an internal or external project. The results serve as a first project briefing and can be complemented to form an initial project business case. The final result of the strategic traceability analysis is a list of operational requirements that are imposed on a traceability system.

2.6.2   System analysis

In the system analysis, the existing production process and traceability information system are modelled in sufficient detail to identify potential bottlenecks. In Chapter 4 (Koenderink and Hulzebos, 2006), detailed instruments for delivering the desired analyses are discussed.

2.6.3   Bottleneck analysis

With the goals for traceability set in the STA, the next step is to describe and analyse the existing system. In two steps, the system is described and bottlenecks in the traceability goals are systematically detected and solved. In Chapter 5 (Koenderink and Hulzebos, 2006), tools for system analysis and bottleneck analysis are discussed. The main focus of system analysis is to describe the existing traceability system. This consists of a description of the processes as well as the information flows and information infrastructure in the chain. In the bottleneck this system description is systematically analysed to detect whether traceability goals that result from the STA can actually be met. The analysis follows the upstream information flow, to check that the required information is available. The lack of information at a certain point in the process is seen as a bottleneck. Iteratively, bottlenecks are identified, and solutions are proposed that may solve the bottlenecks.

2.6.4   System design

The bottleneck analysis leads to changes in the system design to improve or extend the existing system in order to overcome the identified factors that hinder the desired level of traceability. The provided changes are expressed as adaptations on the output of the system analysis. One of the aspects where system design plays a crucial role is in the design of the data model that underlies the traceability system. Especially when traceability becomes a tool to realise added-value functions on top of standard traceability, it is important that the underlying data model is carefully designed to cover these functions.

2.6.5   System construction

With the system design, changes in the system can be realised. There is no standard solution to implement this phase. Steps to take differ on the basis of various factors: is there an existing traceability function that is to be altered, or is it a totally new functionality? Is the existing functionality realised on the basis of a standard solution, or is it tailor-made? Does the repair of the bottlenecks require substantial changes to existing solutions, or does it only require some minor patches?

2.6.6   Implementation in practice

The bottleneck analysis delivers a list of potential bottlenecks that hinder the fulfilment of the traceability goals. Moreover, at the functional design level, solutions are defined to overcome those bottlenecks. In the next step, these functional design solutions are converted into technical designs: low level specifications that describe, in every detail necessary, how the system is built and how it works. The result, a technical system design, is then handed over to system constructors who can actually assemble the working system. As the last step, implementation in practice has to be realised.

In the phases system design, system construction and implementation in practice, FoodPrint is no different to other system development approaches. However, in the design and analysis phase, FoodPrint differs from many other existing approaches.

2.6.7   Alternative approaches

FoodPrint is an approach that has built up from the technical level towards a full-grown system-oriented view on traceability. In many cases, traceability is seen as a specific functionality of other information systems. Typically, traceability is delivered in industrial control systems, MES (manufacturing execution systems) layers or ERP (Enterprise Resource Planning) systems. In all these systems, traceability is approached from specific viewpoints (e.g. process control, logistics). In the FoodPrint view, traceability modules in the mentioned systems can very well be a part of the total solution. FoodPrint offers, on top of these modules, an overall view on traceability that links it to the business goals and that emphasises the strategic role that traceability can play.

2.7   Case studies

When using traceability as a means to optimise business performance, it is crucial to have a clear definition of those business goals. Moreover, it is important to incorporate the business goals into the design phases of the project to deliver. When included in an operational system, the technologies used may not be very spectacular, but the added value of traceability comes from the way they address the specific business goals. In this section, we discuss four projects that we have been involved in, where different traceability technologies served as enabling technology to deliver a business optimisation that may have remained unfeasible otherwise.

2.7.1   Traceability for the organic sector: an example of STA

The organic sector differs from traditional food chains in the application of specific certified production methods and techniques. These methods and techniques are assumed to be more natural than current modern approaches. Organic products are labelled as such and, as a consequence of this labelling, additional revenues are generated. Even for experts it is difficult, and often impossible, to verify on the basis of product characteristics whether or not a product is organic in origin. Consequently, the difference between organic and non-organic products can often only be established through these labels. To guarantee a proper organic origin, a certification authority supervises the organic sector. In return for the right to use the organic labels (EKO in the Netherlands, AB in France, Krav in Sweden, USDA NOP in the USA and a multilingual symbol for the EU according to EEG nr. 2092/91) producers must guarantee organic farming practices. For each of these labels, chain partners such as farmers and processors need to document both origin and production processes of the product. This information, including a traceability requirement, needs to be documented and made available (on request or by default, depending on the certification regime) to the certification authority and chain partners. Recent crises, such as the nitrofen case (2002: nitrofen, a herbicide, in organic grain) and fungicides on raisins, show the necessity for proper traceability tools, for the organic sector as well. If chain partners had been warned at an early stage, the impact of these crises could have been minimised.

The new European traceability requirements were seen as an opportunity to redefine and implement an integrated chain information system for the organic sector. The original idea was to work on two pilot systems according to the FoodPrint STA phasing. Both processes were actually initiated with a first workshop. During these processes, it appeared, however, that key stakeholders had different underlying views on some of the fundamental aspects of chain information systems in the organic sector as a whole. Specifically on the required level of centralism, the basic architecture of a chain information system (central vs. decentralised), an eventual co-ordinating role of a certification body for such a system and the extent to which the perceived system should facilitate other actors (e.g. governments, industries and consumers) generated a lot of debate.

This was the motive to start an STA at the higher sector level. In two workshops, a shared ambition for a chain information system in the organic sector was formulated. The main points of this ambition are that a chain information system for the sector is required to ensure information exchange in the sector. Moreover, the system serves as an instrument to manage product quality, a means to lower the administrative pressure for actors in the chain, a management instrument and as a means for marketing and image enhancement. This STA has resulted in enough confidence to initiate a follow-up project. In this project, the actual definition and design of a sectoral information system is a start.

2.7.2   Traceability for optimised logistics

The production of ornamentals is an important economic sector in the

Netherlands. More stringent demands from clients, combined with the potential of international competition, was the motive for an exporting chain (growers, carriers, exporters) to initiate a project with the goal of optimising the logistic throughput time of the chain. In the initial situation, the chain had a throughput time from moment-of-ordering to moment-of-export of more than a day. Competing chains were supposed to have a throughput time of only a few hours. In an analysis of the logistic organisation eight different scenarios were identified. In all the scenarios, slack time, being the 'waiting for the next step', appeared to be the most time-consuming logistic phase. Moreover, the total process of processing orders had several inefficiencies due to the fragmentation of logistic scenarios. The proposed solution was built up in three parts:

- Designing an improved logistic design to enhance the efficiency of chain logistics
- Introducing a traceability design to enable fast logistics
- Opening up the traceability information for improving business performance (e.g. through benchmarking).

In the project, these results, the logistic design, the traceability system and an internet portal with benchmark results, have been tested as prototype. The concept proved feasible, realistic and useful. It has formed the basis for upgrading the system to (regional) sector level. At this moment, the scaling up of the designed approach is under preparation.

2.7.3 Quality oriented tracking and tracing in the horticultural sector

Quality oriented tracking and tracing (QTT), the use of a tracking and tracing infrastructure for the benefit of quality management in a logistic chain, has received much interest over recent years. The conversion from plain tracking and tracing to QTT requires, amongst other things, a coupling of condition and quality registration to the more traditional registration of product codes (e.g. by scanning bar codes or reading RFID tags). In a two-year project for the Dutch Product Board for Horticulture (Productschap Tuinbouw), the possibilities for applying QTT have been assessed. On the one hand, an overview of available technologies and the consequences for the design of traceability systems have been assessed. Four potential applications have been studied in parallel, in the long-distance export of ornamentals and the trade of strawberries.

Apart from the technical and conceptual results, the project has delivered the pilot applications, in various stages of development. For export of flowers the prototype is an operational internet system for monitoring transport conditions, assessing the quality developments and attributing them to responsible chain actors (E-faqts, 2005). Transport batches are escorted by data loggers that register transport conditions. The registered conditions allow the assessment of the performance of carriers in the logistic chain, as well as the detection of causes for quality problems with the end product. This enables decision making to improve the operational configuration of business processes concerning pre-cooling of products, special packaging arrangements and more accurate prediction of vase life. The system enables batch registration and the downloading of registered conditions. Based on batch data and performance data, various analyses can be made.

2.7.4   Chainfood

Chainfood (www.chainfood.com) is a small innovative company that provides an internet-based traceability approach that realises Chain Quality Management: the opportunity to deploy product information that comes from various chain partners in order to optimise the quality performance of the total chain and of individual chain partners. Chain partners provide their information, either directly or indirectly, to a central data structure (called the Chainfood backbone). This backbone can be used for information exchange between partners, which is valuable in its own. Data providers keep full control and authority over their data, in that they can provide other partners in the chain access to it. Moreover, the company provides additional functions. Current modules include: tracking and tracing (including logistic, product and process information), supply chain management (optimising product and information flows over the total chain) and collaborative business intelligence (the opportunity to monitor and analyse chain-wide patterns in data to unravel useful and previously undiscovered patterns in chain production data).

The data model of the Chainfood system has been designed according to FoodPrint principles at a very generic level. The result is remarkable for two reasons. First, the basic model is limited to a small number of abstract elements, that can be specialised in a specific application. This allows Chainfood to easily adapt the system to new domains. Currently, the system has been customised for pork, dairy and fruit and vegetable applications (http://www.giqs.org/). The aim is to be able to adopt the system in a new domain in a minimum amount of time.

Second, the tracking and tracing module has been developed with an open mind towards the motives for traceability to focus on. Again, the time required to apply the system to a new traceability benefit is very limited. In the generic data model underlying the backbone, the attributes that are required to realise a new traceability benefit can be easily added.

2.7.5   FoodPrint direct

FoodPrint Direct (http://www.FoodPrintdirect.nl, in Dutch) is an internet application that offers a small or medium enterprise (SME) version of the Strategic Traceability Analysis. It has been noted that the STA, as starting point of a FoodPrint project, may be suitable for complex chains and larger companies, but that it is too heavy an instrument for SMEs. To overcome this problem, an on-line version of the STA has been developed as a first screening for SMEs. With this version, SMEs receive a basic analysis on how they comply with legal traceability requirements as formulated in the general food law (GFL) as described in (EU, 2002) and where potential benefits for the company can be expected.

FoodPrint Direct has originally been set up to assist SMEs in preparing the introduction of the GFL. The current version of FoodPrint Direct provides a tool for SMEs for checking their traceability situation and identifying areas for realising added value with traceability.

2.8   Conclusions

The FoodPrint approach is a collection of insights, concepts and tools that have been developed over time. These tools and concepts have been applied in various projects. In this section, some of the strengths and weaknesses of the approach are discussed, the latter resulting in points for further development.

Many companies that start a tracking and tracing development react on external signals. These signals come either from authorities, the General Food Law being an actual example, or from chain partners that require traceability standards to be applied. Typically, in such circumstances, the main focus is on meeting the external demands as fast as possible and not on considering the company's own interest. On top of that it is common to act within a short time horizon, especially in smaller enterprises. It takes some persuasiveness and effort to convince companies to take a step back and consider their own motives for traceability before actually designing a short-term solution.

Related to this problem of commitment for an elementary approach is the problem of speaking the language of the actual workers. FoodPrint is designed to allow a traceability system to be tailored to the specific requirements of food companies, amongst others, in analysing business goals in the strategic traceability analysis (STA). It is crucial to transfer these general principles to the practical situation of the application domain. In our experience, this is a point of FoodPrint that deserves further attention. The experiences with FoodPrint Direct show that there is a genuine interest in more principled traceability approaches, but that accessible communication is a crucial aspect.

It is important to use the commitment mentioned earlier to plan certain stages of the FoodPrint approach in a condensed time frame. For the STA in particular, it is essential to use that momentum. This requires the total throughput time of the STA to be limited (obviously dependent on the number of stakeholders and the complexity of the processes to be analysed).

It is clear that these three points all depend on the quality of the communication with the chain actors. The point of view of the consultant should be close to that of the chain players. It is our experience that FoodPrint is especially suited as an approach for business consultants and system analysts that are close to the daily practice.

A final point that is of crucial importance is to have a detailed knowledge of the applicable legal regulations, standards and certifications. The number of provisions that are applicable to a food company is substantial, varying from general regulations such as Hazard Analysis Critical Control Point (HACCP) to sector or chain specific arrangements such as hygiene codes or the requirements for individual retail chains, with possibly specific requirements on identification, production circumstances, packaging and so forth. As a consequence, it is extremely difficult to extract concrete requirements from the applicable provisions. With the General Food Law being applicable from January 1, 2005, for instance, it was unclear how to interpret the various GFL rules and how and with what criteria the food authorities were going to supervise the proper compliance thereof. Only shortly before the GFL became applicable, the national interpretations were replaced by a European interpretation of norms and a view on supervision and enforcement.

It is our experience that the FoodPrint approach is a useful framework to help companies in defining and designing their traceability system. The fact that traceability is more than an enforced set of rules and that the implementation can actually help to realise specific business goals is an eye-opener. It helps companies to come up with a balanced solution for traceability. This solution will often be based on standard solutions. In those cases, FoodPrint helps to formulate selection criteria. In other cases, the solution is tailor made, and reflects the specific requirements of the company.

Some of the FoodPrint concepts, especially in the phases of system design and system construction, deserve further attention. Based on experience in a number of projects, the tools and concepts will be further elaborated. The basic innovation of the approach, however, lies in the initial phases.

In recent years, food traceability has become a standard condition in production and distribution chains. Food production and trading will soon be impossible without having a basic traceability system operational. This means that traceability will not lead to additional income for food producers and traders. There will not be additional revenues in return for the investments that companies are forced to make.

The only way companies can benefit from the investments in traceability is by generating their own return on traceability investments. In this chapter, we have argued that the potential benefits lie in improved cost control and adding value to products. The natural reaction of companies is to cut costs. This is indeed a sensible first reaction and cutting costs always pays back. However, it is challenging to identify and realise options for adding value to products.

Methodological frameworks that help companies to identify and realise options for cost saving and realising added product value in combination with traceability will therefore, in the long run, be more beneficial for companies to deploy. One of the approaches that realises this options is FoodPrint.

2.9   References

Bowersox, D J, Closs, D J and Stant (1999), T P, *21st Century Logistics; Making Supply Chain Integration a Reality*, Chicago, Council of Logistics Management, Ch. 10.

E-faqts (2005), http://www.agrotechnologyandfood.wur.nl/nl/Producten+en+Faciliteiten/E-faqts.htm, version of February 2005.

EU (2002), Regulation (EC) No 178/2002 of the European Parliament and of the Council of 28 January 2002, http://europa.eu.int/eurlex/pri/en/oj/dat/2002/l_031/l_03120020201en00010024.pdf, version of February 2005.

FoodTrace (2005), Compendium of Data Carriers for Traceability, deliverable of FoodTrace, EU Concerted Action Project QTLRT-2000-02202, http://www.eufoodtrace.org/files/compendium.pdf, version February 2005.

Het Financiële Dagblad (2002), Label on meat does not provide certainty (in Dutch), 7 October.

Imtech (2004), *Food Management*, September.

Koenderink, N and Hulzebos, L (2006), Dealing with bottlenecks in traceability systems, in: Smith, I and Furness, T, *Improving traceability in food processing and distribution*, Woodhead Publishing, Ch. 5.

Koenderink, N and Hulzebos, L (2006), Modelling food supply chains for tracking and tracing in: Smith, I and Furness, T, *Improving traceability in food processing and distribution*, Woodhead Publishing, Ch. 4.

Pape, W R, Larson, D and Jorgenson, B (2002), Traceability – Cost Burden or Profit Opportunity? *Food Traceability Report*, May 2002.

Sloof, M, Tijskens, L M M and Wilkinson, E C (1996), Concepts for modelling the quality of perishable products, *T. Food Sci. Technol.*, **7**, 165–171.

Vernède, R and Wienk, I (2006), Storing and transmitting traceability data across the food supply chain in: Smith, I and Furness, T, *Improving traceability in food processing and distribution*, Woodhead Publishing, Ch. 10.

3

Optimising supply chains using traceability systems

F.-P. Scheer, Wageningen University and Research Centre, The Netherlands

3.1 Goals and benefits with quality-oriented tracking and tracing systems

Traceability becomes compulsory under the General Food Law from January 2005. However, complying with this law does not offer added value or reduced costs for companies in the food chain. It only offers a licence to produce. Obligations by law alone are therefore not a good motivator for food companies to invest in tracking and tracing (T&T) systems. A commercial-cum-strategic approach is needed to obtain more benefits. Having identified these benefits, the offset with the required investments in such a T&T system improves. The identification of the goals is the starting point for a quality-oriented tracking and tracing system (QTT).

With a conventional T&T system, the vast amount of collected data about location and identification of products is usually only used for recall management. With a QTT system, data about relevant parameters for product quality, for example, temperature and relative humidity throughout the chain (chain conditions), are also captured. The combined T&T and quality data are continuously used (active approach) to control and manage the flow of products in the food chain, resulting in more benefits.

3.1.1 Benefits with QTT

A QTT system offers benefits, especially for fresh, perishable products like vegetables, fruit, meat, fish, dairy products and flowers.

First, a QTT system offers an improved assurance for food safety. By knowing the food quality throughout the chain, it is possible to take unsafe food out of this chain before it reaches the consumers. Another way is to adjust the surrounding conditions of the food product in such a way that the required quality standard is met and unsafe products are avoided. This first category is the so-called 'compliance-oriented strategy'.

Second, a QTT system offers an improved customer service level. It becomes possible to optimise towards specific quality demands from customers, for example a specific colour, ripeness, texture, taste, grade or shelf-life. Mangoes, for example, are often unripe and raw (hard texture) when they are offered in retail shelves, while most consumers search for 'ready to eat' mangoes. This second category is the so-called 'market-oriented strategy'.

Third, with QTT, costs can be saved, for example, by reducing product loss. Product loss of fresh products is one of the most important aspects in retail, due to high costs of wasted products when due dates are passed. To avoid product loss, demand and supply chain management of fresh products is far more critical than for dry foodware. Critical for fresh products is to manage the dual goal of avoiding out-of-stock (no sales) on the one hand and reduction of product loss (reduced costs) on the other, due to their limited shelf-lives. This third category is the so-called 'process improvement-oriented strategy'.

Fourth, with a complete and accurate QTT system, the logistical benefits of conventional T&T (identification and location) are also better utilised in, for example:

- Improved service through reduced out-of-stock levels.
- Lower (safety) stock levels by improved demand and supply chain management.
- The number of re-deliveries (corrective) can be reduced.
- The loading level of trucks can be increased.
- Savings in time and personnel by automatic identification and registration (AUTO-ID). Auto-ID also results in fewer mistakes and more on-line insight into the actual situation (stock levels etc.).
- Improved utilisation of resources (e.g. space in distribution centres (DCs) and trucks).
- More efficient retour logistics of pallets, dollies and crates.

By identifying all requirements before investing in a T&T-system, more benefits can be obtained with QTT. Therefore, this proactive approach is completely different from a reactive implementation of T&T for legal and food safety purposes only. The proactive QTT approach results in an improved payback when investing in a tracking and tracing system. Traditionally, the food industry puts up a lot of resistance complying with T&T requirements, because it adds costs to the bottomline and does not offer enough benefits in return. QTT does offer more benefits and is therefore a better motivator for food chain actors to invest in such a system. The process of determining QTT benefits is described in Chapter 2.

3.1.2 QTT system requirements

The elements of a complete QTT system are summarised in Fig. 3.1. Later in this chapter, the different elements will be described in more detail.

1. Goal identification and benefits for tracking and tracing. This element was covered in section 3.1.1.
2. Demand and supply chain management (DSCM). In order to achieve the goals, how are demand and supply in a food chain managed (volume and quality) and who in the (chain) organisation should be responsible for a certain task? DSCM, specifically for fresh perishables is the main focus of this chapter.
3. Information: Which QTT data are needed to manage and organise the chain, at what level (crate, pallet and container) and at which sampling frequency? The way quality information is used will be described in this section.
4. Technology: Which technologies (barcodes, RFID, sensors, etc.) can be deployed to register the required quality and T&T data. This element is covered in Chapter 10.
5. Product and process: What is the product and what are its characteristics? For example, strawberries are much more susceptible to quality decay than apples. Process identifies the chain links, such as the market that is supplied by multiple suppliers using different transport modalities (truck, boat, plane).

The five system elements are all dependent on and related to one another. As shown in Fig. 3.1, layer 3 (information) and layer 4 (technology) are the supporting elements for providing the required information to manage the food chain (layer 2). The system is product and chain specific (layer 5). Demand and supply chain management (DSCM) of perishables is the management 'layer' of a QTT system and is the main focus of this chapter.

Quality decay models are important tools in QTT systems and are part of

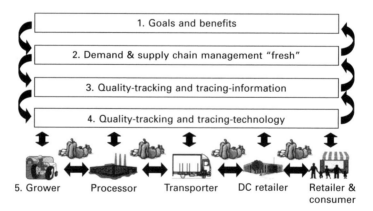

Fig. 3.1 The elements of a quality-oriented tracking and tracing system.

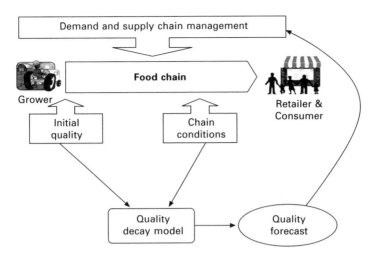

Fig. 3.2 Demand and supply chain management based on product quality.

the information layer (3). The models translate obtained quality data (layers 4 and 5) into required management information (layer 2), for example, remaining shelf-life. The quality decay model has two main inputs. First, initial quality at harvest and, second, the chain conditions (such as temperature and relative humidity over time). In this way, remaining shelf-life throughout the chain can be forecast and acted upon in the food chain from farm to fork (Fig. 3.2) With this information, demand and supply chain management of fresh food products aims for optimal chain control.

3.2 Demand and supply chain management

Demand and supply chain management can be defined as follows:

> Demand and Supply Chain Management (DSCM) is the management of a network that links customers and suppliers as one 'single entity' with the objectives to create value and reduce waste through the voluntary integration and coordination of the objectives of three or more – and, ideally, all the – independent parties in the network[1].

From this definition, it is clear that DSCM and therefore QTT is a chain-wide approach aiming for benefits for all chain partners. The definition also identifies multiple goals, creating value (market strategy) and reduction of waste (process improvement strategy). We add food safety (compliance strategy) to this definition as an important goal in food chains.

As in most chains, the food chain is also a typical 'demand and supply' managed chain. Typically, the market demand and the consumer's buying behaviour are the starting points of product sales, followed by supply from the supplying chain actors. DSCM is not specific for food chains; however, the management of demand and supply is more critical in fresh perishable food chains than for dry-foodware. Demand and supply have to be tied very closely together due to the limited shelf-life of the food products. Traditionally, undersupply will result in out-of-stock, while oversupply will result in product loss. A fine balance, therefore, has to be found to manage both low out-of-stock levels and at the same time low product loss levels. This last aspect of balancing out-of-stock with product loss is typical for DSCM in food chains. As product loss is such an important topic in food chains, we will focus more deeply on this aspect in the following sections. By doing this, we will show what the additional benefits of a QTT system can be when this aspect is incorporated in the system. Other improvements mentioned in section 3.1, such as improved market service, are not described in this chapter, but do, however, offer similar opportunities for improvement as for the case of product loss.

3.3   Product loss and out-of-stock levels

We define product loss as 'value loss due to quality decay'. Other stock losses such as theft or damages are therefore not taken into consideration in this approach as they are not quality related. Product loss usually represents 80% of stock loss and, thus, is the most important aspect of stock loss.

As this aspect of product loss is typical for fresh food, it does not occur for dry foodstuffs (e.g. coffee or rice, which have long shelf-lives) or for other consumer products (like computers, TVs etc.). Oversupply of non-fresh products does not result in product loss; however, there can be stock costs and commercial loss when newer versions of a product are introduced. We will also not take these commercial losses into consideration.

Product loss does not only occur on the retail shelf. When the remaining shelf life in the supply chain is not sufficient, product loss also occurs there, for example, at the site of a processor, trader, auction or transport company etc. Therefore the reduction of product loss offers benefits for the entire chain. When combining numbers out of several A&F food research projects, the following overview is obtained (see Table 3.1)

From Table 3.1 it can be concluded that product loss is a huge problem in food chains and it is an important aspect to act upon. Table 3.1 presents average product loss levels in food chains. Products like mixed vegetables and organic meat show even higher percentages. Contrary to this, products like milk, tomatoes and apples show lower percentages. Later in this chapter we will discuss the main reasons for these differences.

On the other hand, out-of-stock has been a traditional problem. A worldwide examination of extent, causes and consumer response was done in 2002 by

Table 3.1   Average percentages of product loss for fresh products per chain actor and in total

Chain actor	Average product loss, %
Total chain	35
Consumers	15
Retail	5
Processing and distribution	5
Primary production	10

the universities of Colorado, Emory and St. Gallen[2]. Some of their findings were:

- The overall out-of-stock (OOS) rate is estimated at 8.3%.
- The majority of out-of-stock tends to fall between 5 and 10%. Exceptions are for, e.g., promotions which tend to exceed 10%. Hair care products have the highest out-of-stock levels with an average of 9.8% (varies between 7.0% and 16.0%)
- 70–75% of out-of-stock is a direct result of retail store practices. Causes are 28% upstream, 25% in the store other than on the shelves and 47% due to store ordering and forecasting.
- Consumer responses to out-of-stock situations are: do not purchase item (9%), substitute a different brand (26%), buy item at another store (31%), get substitute from the same brand (19%), delay purchase (15%).

Although there are differences between countries and products and this survey was not specifically for perishables, it can be concluded that out-of-stock is also a serious aspect which should be acted upon. As 70–75% is retail related, it means that most of the attention should be given at the retail store. As described, a fine balance has to be found to manage both low out-of-stock levels and at the same time low product loss levels. Financially product loss has more impact than out-of-stock. Product loss namely represents the full cost-price of a product (100%). Out-of-stock results in lost margin (typically 25%) due to lost sales. As product loss represents a 100% loss and out-of-stock is only 25%, reduction in product loss therefor offers more financial benefits than reduction in out-of-stock. However, it must be mentioned that empty shelves have a negative effect on service and sales volume. These are harder to quantify.

3.4   Causes of product loss and out-of-stock

Based on the demand and supply chain management (DSCM) methodology, the causes of product loss and out-of-stock may be divided into two groups:

demand-driven and supply-driven causes. First, we describe the main variables from the demand-side:

- Turnover: the turnover of a product is variable and is related to many different aspects, such as:
  - Demographic profile: differences due to income-level, social and cultural preferences, location (city versus country), etc. result in different buying behaviours.
  - Weather circumstances: products like ice creams and barbecue meats have increased sales during summer periods.

  Although variable sales are a fact and do result in either oversupply (product loss) and/or undersupply (out-of-stock), sales volumes can be forecast to a certain level. Accurate data capturing at the point of sale and a fast response of the supply chain (short leadtimes) are essential. In terms of percentage of variance in sales of a high-turnover product, this tends to be lower than for a low-turnover product. As a result, high-turnover products can be better managed and give lower levels of product loss. However, in terms of percentage values, product loss and out-of-stock for low-turnover products are lower and the absolute financial loss can be as big as or bigger than that of high-turnover products. This is because the volume sold is much greater for high-turnover products. It is therefore essential to find the optimum of low product loss levels and out-of-stock levels to reduce financial losses as much as possible. Order management at the retail store is a key operation to achieve this.
- Order management: to avoid out-of-stock and thereby lost sales, sufficient and in-time deliveries have to take place. The supply of the right amount of products at the right time and the right place is a response of the supply chain to the market demand by consumers. Order management is a critical task because the right amounts have to be ordered in the face of much uncertainty about turnover and customer behaviour. Bad order management is therefore in itself a reason for higher levels of product loss and out-of-stock.
- Retail formula: the chosen formula can result on the one hand in a small assortment, low cost, low service formula ('price-fighter') and, on the other hand, in a wide assortment, higher price, higher service formula (service retailer), while in between there are a range of possibilities. The number of SKUs (Stock Keeping Units) offered and the required service level has a direct impact on out-of-stock and product loss levels. Therefore, a certain accepted level of out-of-stock and product loss are connected to the chosen formula.
- Customer behaviour: what is the customer response to price deduction, number of facings, shelf position and so forth? This is often hard to define as there is not such a thing as THE customer. The best approach is to identify customer profiles with specific shopping behaviour. Moreover, what is the picking behaviour when multiple due dates are available for one product? Picking behaviour is typical for fresh products with limited

shelf-lives. For fresh products, there is not a typical FIFO (First In First Out) picking, because most customers demand the longest shelf-life so they act as LIFO (Last In First Out) customer. This picking behaviour results in extra product loss because products with the shortest shelf-life are left behind on the shelf whilst products with the longest shelf-life have been picked by the consumers. LIFO picking and product loss can be partly compensated by price deductions close to the due date. The latter is called price-related product loss because products are sold at a lower price.

The supply-driven causes of product loss and out-of-stock are as follows:

- Supply chain management: the supplying chain actors will respond to the orders as placed by different retailer stores. Each chain actor has to balance production capacity and safety stocks to avoid undersupply (out-of-stock) and to avoid high stocks (product loss). Due to limited shelf-life of food products, demand and supply have to be tied very closely together. Supply chain actors therefore have to find a similar balance between out-of-stock and product loss. In this way, the storage room of a supply chain actor (producer, auction etc.) acts like a retail store.

- Shelf-life: the longer the shelf-life, the more selling days are available before the due date of the product is passed, resulting in waste streams and product loss. Control of shelf-life is, therefore, very important for perishable products. The main parameters contributing to shelf-life are:

  - Initial quality of the product at harvest. First, there are big differences among product groups; for example, strawberries with a limited shelf-life of 2 to 3 days compared with apples with a shelf-life of a few weeks. Within a product group, there are big differences again. This can be a result of different varieties but the growing circumstances also have large effects. Besides the environmental conditions such as weather and soil type, one should think of the application of chemical treatment and fertilisers. Biological variance causes varying responses of individual food items to these variables, resulting in differences in initial quality. To cope with this aspect, it is important to separate homogeneous batches as much as possible. When this is done, it becomes possible to act upon these differences by the following actors of the food chain.

  - Processing: there are several processing techniques available to extend shelf-life, e.g. pasteurisation, near infra red and high-pressure processing.

  - Chain conditions like temperature and relative humidity: most products have a preferred upper and lower limit of chain conditions. When these limits are exceeded, quality decay is the result. Examples of these are bruising, softening or drying out.

  - Distribution time as a result of transport distance and modality (boat, truck, train and airplane): the combination of time and temperature is called 'temperature-sum' and is an important parameter to forecast shelf-life.

  - Packaging of the product: Modified Atmosphere Packaging (MAP) is a way to extend shelf-life by adding a certain gas atmosphere around the product. The gases are usually a combination of carbon dioxide (to limit bacterial growth), oxygen (to retain a fresh colour) and nitrogen (ambient air). By applying MAP, the shelf-life usually increases by at least 50%, for example from 4 to 6 days shelf-life.

3.5   Measures to control product loss and out-of-stock

A&F Innovations has developed several tools and solutions to manage product loss and out-of-stock in the most appropriate ways.

3.5.1   Product loss monitor (PLOMON)

The first step is to identify the levels of product loss and out-of-stock. Preferably this analysis should be done specifically per retail formula, per product category, per product, per store, per day and trends should be available over time. To do this, the 'product loss monitor' has been developed. The monitor is fed by point-of-sale data translated into required information that can be acted upon. An example of this is the insight into how product loss is divided over different products and stores so priorities can be identified. It is also possible to evaluate a certain product category on its total turnover and revenues (category management).

3.5.2   Shelf-ordering management system (SOMS)

Based on point-of-sale data and the insight gathered by the product loss monitor, specific order rules are advised by the shelf-ordering management system (SOMS). An important aspect is what the safety stock level of a product should be given, its sales pattern and balancing out low product loss and out-of-stock levels at the same time. To do this, SOMS imports the daily sales and stock levels and exports an ordering advice specific per product-day and store. An ordering rule is created by statistical analysis of the sales pattern. The system is self-learning because orders of day 1 are automatically evaluated at day 2 (and so forth) by the product loss monitor.

3.5.3   Techniques for extended shelf-life (TECHNES)

Extension of shelf-life is very important to extend the time between delivery and sales. Because sales are variable by nature, a longer shelf-life gives more sales time to result in lower product loss levels. To extend shelf-life, several options are available. As mentioned, the main variables for shelf-life are initial quality, processing, chain conditions, distribution time, and packaging. For each of these variables, techniques are available to extend shelf-life.

3.5.4 Shelf-life monitor (SLIMON)

The shelf-life monitor (SLIMON) evaluates different product-market combinations on shelf-life. SLIMON uses the earlier mentioned quality decay model to translate quality data into shelf-life information. The performances of different chains on shelf-life can be compared and the effect of improvements (for example the usage of modified atmosphere packaging) can be measured.

3.5.5 Predicting initial quality

The initial quality at harvest is an important starting point. In the current situation, however, no direct measurement methods are available to objectively quantify the product's initial quality. There are only many indirect indicators. One has to think of product, variety and growing circumstances (weather, soil type, chemical treatment, fertilisers). When data of all these indicators are combined in a so-called 'learning system', trends between pre-harvest indicators and post-harvest quality can be derived. When these learning systems are combined with differentiation techniques at harvest (like camera vision), more homogeneous batches can be forwarded into the supply chain.

3.5.6 FEFO control

To reach the benefits which were mentioned and to reduce product loss in particular, it is essential to know the remaining shelf-life of products during distribution throughout the chain, so that action can be taken. FEFO is the abbreviation of First Expired First Out, meaning products with the least shelf-life are distributed first. Contrary to the conventionally used concept of FIFO (First In First Out), FEFO acts upon remaining shelf-lives. By utilising FEFO control in the food chain, product loss levels can be reduced by several tens of per cents. In some cases, 50% product loss reduction has been achieved. When, for example, product loss has been reduced from 10 to 5%, a cost reduction of 5% can be achieved, because product loss represents the full cost price of a product. In this example, 5 billion euros yearly have been saved for every 100 billion euros of turnover. This shows it is very effective to change the logistical control to FEFO. Examples of various control measures as a result of FEFO are:

- Changing from local to international distribution (longer distribution time) when shelf-lives are longer than expected and vice versa.
- Matching shelf-life with turnover of individual retail stores. Longer shelf-life is used to decrease product loss in low turnover outlets. Products with shorter shelf-life are sent to high turnover outlets, since product loss will not increase for these outlets due to the high turnovers (products are sold before reaching due-date). The average result of low- and high-turnover outlets for the total retail formula is a decrease of product loss.

3.6   Demand and supply chain management optimisation

Figure 3.3 shows a visual overview of the demand and supply chain management cycle. In this case, the DSCM of the retailer is chosen, but this cycle is also applicable for the other food chain actors.

1. Shelf-management starts with the consumer's buying behaviour, resulting in point-of-sale (POS) data.
2. The product loss monitor PLOMON translates the raw POS data into information of product losses and out-of-stock. Specific reports generate insight on product and store level, which can be acted upon by the category manager and store order managers.
3. The shelf-order management system SOMS translates the raw POS data into specific orders per product. A balance has to be struck between low out-of-stock levels and low product loss levels. The SOMS system can be seen as a decision support system, advising the store order manager. The store order manager is advised by SOMS but can overrule it if there are good reasons to do so, for example from information on promotions or knowledge about the local buying behaviour. In another configuration, SOMS generates all orders for all products of all retail stores. In this way, SOMS generates automatic replenishment orders to the supplying parties upstream in the supply chain.
4. Suppliers will replenish with the right amount and the right quality for the right customer at the right time. Different techniques are available to extend shelf-life (TECHNES).
5. The shelf-life monitor SLIMON can evaluate current shelf-lives and the effect of shelf-life extension techniques.
6. As there are many variables that have their effect on shelf-life, differences in shelf-life occur. With FEFO control, the logistical control reacts on

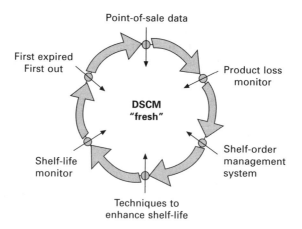

Fig. 3.3   The demand and supply chain management cycle for fresh food products.

these shelf-life differences in a smart way to reduce product loss and out-of-stock levels as much as possible.

When sales of day 1 have passed, the cycle repeats itself on day 2. Actual sales reports of the previous day become available (per store, product and day). PLOMON calculates key performance indicators and SOMS generates orders based on past experience and future forecast, and so forth.

3.6.1 Supply chain optimiser ALADIN

ALADIN stands for Agro Logistical Analysis and Design Instrument and is a supply chain simulating and optimising system. ALADIN is specifically developed for perishable products. ALADIN supports (strategic) decision making by simulating different scenarios, by calculating remaining shelf-life and by reporting costs and benefits. For example, what is the effect on product loss and out-of-stock when distribution is changed from airplane to truck; what is the effect of pre-cooling or what is the effect of modified atmosphere packaging? Similar to the real world, sales data are imported into the system. Together with the effects on shelf-life, ALADIN reports the costs of an investment (for example pre-cooling, map) versus the benefits (product loss and out-of-stock) for each scenario. The results are both actor-specific as for the total chain. By using ALADIN as a diagnostic instrument, the optimal chain configuration can be found before big investments are made in the real world. The latter often incurs high cost for trial and error.

3.7 Conclusions

A&F considers tools and techniques as means to reach a specified goal. Tools and techniques are not a goal by themselves but are specially designed to fulfil a certain task that can help companies in the food chain to achieve better results in a more sustainable way. QTT has been developed over a period of years. As the concept has existed for years, its application and comprehension level is constantly updated to a state-of-the-art level with the newest insights. In particular, the interaction between techniques (beta-science) and chain management (gamma-science) has proven to be very effective. Techniques offer ways for improvement. Chain management allows these techniques to be used in such a way that they offer real added value and can be practically applied in real situations.

Despite the state-of-the-art level of current QTT systems, opportunities for improvement are always there. As product quality of perishable foods is variable by nature, the objective and accurate registration of its value and forecast of its future development remains an everchallenging topic. Like the discovery and description of the human genes, a lot of the product quality is still unknown. Besides these technical aspects, ethical aspects play a role too, in judging whether full knowledge of the natural products is to be recommended.

3.8   References

1. Ploos van Amstel, W and van Goor, A R, *Van logistiek naar supply chain management*, The Netherlands, THS Publishers, 2001.
2. Gruen, T W, Corsten, D and Bharadwaj, S, '*Retail Out-of-Stocks: a worldwide examination of extent, causes and consumer responses*', Washington DC, Grocery Manufacturers of America, 2002.

Part II

Building traceability systems

4

Modelling food supply chains for tracking and tracing

L. Hulzebos and N. Koenderink, Wageningen University and Research Centre, The Netherlands

4.1 Introduction

Companies in food supply chains are confronted with legislation requiring them to deploy an adequate tracking and tracing (T&T) system. A tracking and tracing system contributes to food safety and detects liable parties in case of unacceptable food risks. The implementation of a tracking and tracing system can also be used simultaneously as an opportunity to implement strategic, tactical and operational goals (see Chapter 2).

Tracking and tracing related projects inspired us to develop the FoodPrint method. FoodPrint is a structured methodology to create a goal-oriented tracking and tracing (T&T) system that builds on the operational and strategic goals of the production network. The method consists of the following steps:

- **Strategic traceability analysis**: making an inventory of the strategic goals derived from operational goals (see Chapter 2);
- **Traceability system analysis**: charting the current tracking and tracing system;
- **Traceability bottleneck analysis**: checking whether the current information situation suits the operational goals. Any discrepancy leads to adjustments in the tracking and tracing system (see Chapter 5);
- **Traceability system design**: leading to a proposed design for the new tracking and tracing system. After this stage, the implementation of the tracking and tracing system takes place.

There is a general overview of the FoodPrint method in Chapter 2.

In this chapter, we present a systematic modelling approach that can be used to analyse the food supply chain and the characteristics of the required tracking and tracing system are presented. This approach results in two models:

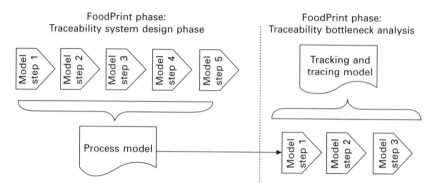

Fig. 4.1   Generated models during the FoodPrint method phases and its related modelling steps.

the *process model* and the *tracking and tracing model*. The process model is created in the Traceability System Analysis stage of the FoodPrint method and yields an accurate representation of the considered food supply chain. Next, he bottleneck analysis stage transforms the process model into a tracking and tracing model. This tracking and tracing model contains all concepts and relations needed to carry out a thorough bottleneck analysis. The (adjusted) models are input for the Traceability System Design where the tracking and tracing system that is to be created or adjusted is specified, together with the necessary adjustments in the food supply chain (as described in Fig. 4.1). The concepts that are the building-blocks of the models were developed as a follow-up of a co-operative generic chain modelling project between Agrotechnology & Food Innovations and Chainfood (Chainfood, 2004).

Throughout this chapter, we will use the example of an industrial facility for processing fresh tomatoes. In this facility, fresh tomatoes from tomato farmers are delivered. After arrival, the tomatoes undergo a quality check by quality controllers; the approved batches of tomatoes are successively peeled and canned. Finally, the canned tomatoes leave the processing facility. After the introduction of the models we will discuss and address important modelling issues regarding ident mapping and ident resolution in food supply chains. We end this chapter with some future trends in tracking and tracing modelling and concluding remarks.

4.2   Developing a process model

To design a proper T&T approach we have searched the literature for available modelling techniques to chart a food supply chain. No modelling technique we discovered covers process and product registrations and the relationships we consider to be relevant for T&T modelling. Therefore, we developed a dedicated process model which focuses on T&T and is applicable to any given (food) chain.

This process model describes a food supply chain with respect to the registration of products, measurements on products, measurements related to processes and events related to processes. These relations are essential to reproduce the information needed for a tracing activity. There are five steps to be taken when making a process model. Each step leads to the introduction of some concepts used in the process model.

4.2.1 Defining the scope

First, it is necessary to define the scope of the model and this follows from the FoodPrint Strategic Traceability Analysis. Within the FoodPrint method, a chain is a collection of companies or organisation divisions sharing a common interest in producing, shipping or manufacturing a product or set of products. It is assumed that the members of the chain are willing to share tracking and tracing information. For example, we can imagine ten tomato farmers, a peeled tomato factory and five retailers who are willing to share T&T information. Those tomato farmers who are not able or willing to share T&T information are seen as suppliers to the chain, and those retailers who are not willing or able to share T&T information are seen as vendors of the chain (see Fig. 4.2).

A chain consists of chain links, i.e. the participating companies. A chain link is a value-adding activity and a legal entity providing products or services to the chain. Recalling our example, the factory 'Wageningen Peeled Tomatoes Inc.' is such a chain link. We assume that, within a chain, the entities trust each other and are willing to share T&T information. So the issue of trust, even though it is very important, is not considered here. For the sake of clarity, the entire chain in the example is not modelled, and the scope of the process model is restricted to the peeled tomato factory. This process model is depicted in detail in Fig. 4.3.

4.2.2 Identifying the processes within the scope

Within a chain link, value adding activities take place, such as, in our example, the peeling of tomatoes in the tomato factory. There is a hierarchy in those activities. On the top level, the peeled tomato factory does 'produce peeled tomatoes'. This activity, or process to be more precise, comprises 'intake of tomatoes', 'sorting tomatoes' 'peeling tomatoes' and 'canning tomatoes'. When giving 'sorting tomatoes' a closer look we can break this process down into the subprocesses 'sort tomatoes by size' followed by 'sort tomatoes by colour'. This level of description can be further broken down into subprocesses, until an elementary description level is reached (see Fig. 4.3). Within the peeled tomato factory, four processes are identified: P1 'Intake of tomatoes', P2 'Outtake of tomatoes, P3 'Peeling tomatoes', P4 'Canning tomatoes'. The modeller decides that this level of detail suffices and no further breakdown is needed.

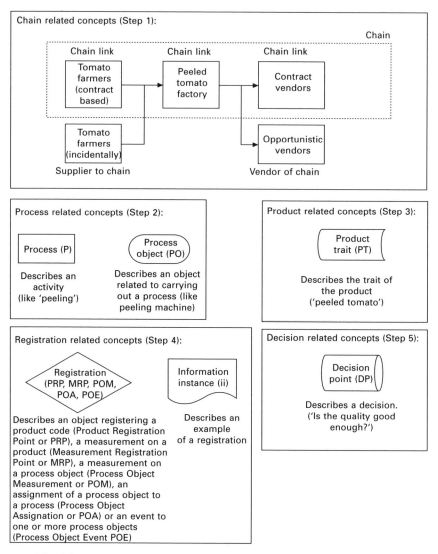

Fig. 4.2   Concepts used in the five steps towards creating a process model.

Processes are carried out or supported by one or more 'objects'. An operator controls the mixing process, while a mixer is doing the actual action. In this case, the operator and the mixer can be represented as process objects, or PO for short. In the same sense, a warehouse can be represented as a process object related to the process 'storage'. The process objects can participate in multiple processes. For example, employee Pete supervises both the process 'mixing' and the real world process 'sorting'. Process objects can be grouped into collections of process objects. So the process objects John and Pete can be grouped into the collection 'line operators'. The advantage of using 'line

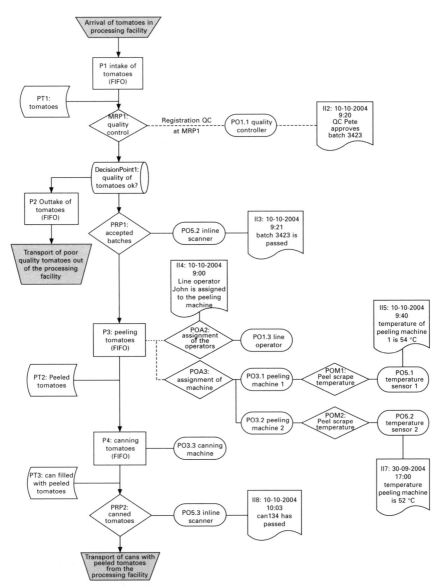

Fig. 4.3   Process model of production of peeled tomatoes in a processing facility; the diamonds indicate a registration.

operator' instead of 'Pete' in the process model is that not every operator's name has to be written down (see Fig. 4.4). The collection representing all process objects (PO ALL) is decomposed into the categories PO1 for resource, PO2 for information system, PO3 for production inventory, PO4 for administrative subjects and PO5 for measurement devices. PO1 is further decomposed into PO1.1 for quality controller, PO1.2 for T&T operator and PO 1.3 for line operator.

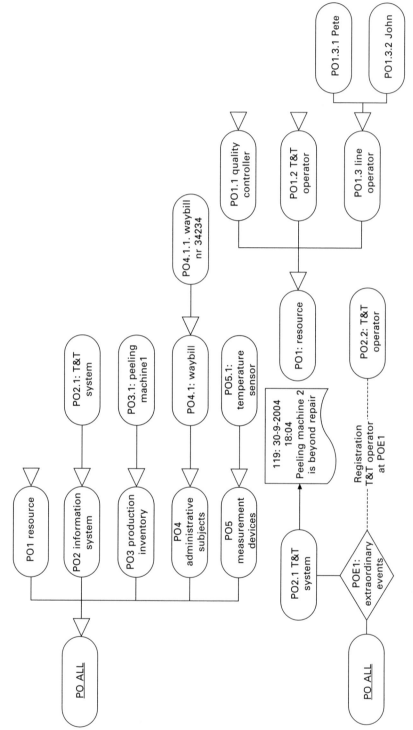

Fig 4.4 Relationships between process objects; for convenience the process objects are clustered hierarchically.

4.2.3   Capturing the product

For T&T purposes, the products are the main subjects of concern. When tracking it is essential to know the current whereabouts of the product in a food chain. Therefore registrations of the product have to be made. A product like a can of tomatoes may contain one or more product codes like a barcode, a printed batch code, or an electronic RFID code. These codes are registered by a T&T system (i.e. on a piece of paper or more likely in an information system). To separate the technology from the idea we consider that registrations are made in information space (for example by retrieving an electronic code by a scanner). That electronic code in the T&T system is what we call an ident. It represents the identification of the physical object in the information space (see Fig. 4.3). Ident Can134 is the representation in information space of a particular can filled with peeled tomatoes.

A product that travels through the chain may undergo a change of trait. For example, considering the products in the peeled-tomato chain: the product at one point is a fresh tomato, at another point a peeled tomato and, finally, a can filled with peeled tomatoes. We record these product traits together with the size of a batch, since they form an important indicator for tracing activities (see Fig. 4.3). There are three different traits of products spotted: PA1, fresh tomatoes, PA2: peeled tomatoes, PA3: can filled with peeled tomatoes.

4.2.4   Capturing the registrations

Two types of product registration are distinguished. A registration that is solely made to record the location of a product is called a product registration. Such a registration is made at a product registration point (PRP). When a product is registered and an additional measurement is performed, we call the registration point a measurement registration point (MRP). MRPs can also be off-line measurements like a laboratory test on a tomato sample which takes a week to generate results (see Fig. 4.3). The quality control of the incoming tomato batch is represented as a measurement registration point, namely MRP1 'quality control'.

The status of a process object may change over time and this may be important information for the (quality-oriented) T&T goals of the chain. It could, for example, be important to monitor the temperature of a warehouse during the day. We call such a measurement a process object measurement (POM). Note that another process object is carrying out the actual process object measurement (see Fig. 4.3). Process object PO5.1 'temperature sensor 1' carries out the actual measuring of POM1 'Peel scrape temperature'.

The relationship of a process object to a process can change over time. Figure 4.3 shows that peeling machine 1 and peeling machine 2 can carry out activities of process P3. It is possible that, during a particular period, peeling machine 1 is active, whilst peeling machine 2 is in maintenance. The relationship between a process object and a process can be crucial if it turns out that peeling machine 1 was contaminated for a certain time period and peeling machine 2 was not. The mechanism that describes changes of relations between process objects and processes is called a Process Object Assignation (POA) (see Fig. 4.3). POA3 'assignation of machine' registers whether peeling machine 1 or peeling machine 2 was deployed in process P3 'peeling tomatoes'.

Irregular but, for T&T, relevant events like a power shortage are different from process object measurements and may affect one or more process objects. A power failure may lead to less reliable measurements. These faulty readings may implicate a quality risk. We register such events with a concept called a Process Object Event (POE). Since those irregular events can be diverse, the registered variable is not fixed (in contradiction to a POM, that measures the value of a predefined variable in a certain unit (e.g. the temperature is 10 °C). In Fig. 4.3 a Process Object Event (POE1) is assigned to the aggregated process object collection named ALL. This means that any irregular event to any process object is registered in the T&T system by POE1 (see Fig. 4.4). A process Object Event is represented by POE1 where the incident is registered that peeling machine 2 is beyond repair.

To know what information is actually available at a certain registration (at a PRP, MRP, POM, POA or POE) we introduce the concept of Information Instance (II). Such an II represents an example of a certain registration (see Fig. 4.3). We see that II3 represents a detection of tomato batch 3423 on 10-10-2004 at 9:21 (24h notation).

4.2.5   Capturing the decisions

A decision point is a process object, i.e. machine or operator that reads the product identification and is able to access all the relevant information about that particular product in the T&T system. Since only data is read from the T&T system, a decision point is not seen as a type of registration. The decision point decides on the outgoing flow of the product. In Fig. 4.3 we see that based on the known information about the passing tomatoes, Decision Point 1 makes a decision about the quality of tomatoes. If the decision is negative, the tomatoes are removed from the chain. If the decision is positive, PRP2 'accepted batches' makes sure that the approved tomato batches are registered into the T&T system.

4.3   Creating a tracking and tracing model

The T&T model reflects all registrations in the T&T system regarding T&T and the relationships of these registrations towards processes and products. Those relationships make it possible to trace upstream or downstream through the chain. It addresses questions such as 'How do I link a can of peeled tomatoes to the batch (or batches) of tomatoes that are inside the can and vice versa?' The T&T model focuses solely on the available information on the product, rather than on the physical process itself. To be able to classify the different kinds of information links between the process steps, the concept of relation type is introduced into the T&T model.

4.3.1   Drawing the product-related registrations

In the T&T model, the registration points play an important role. All registration points (the PRPs and MRPs) in the process model reappear in the T&T model. To focus on the relationships between products and registrations, we keep the T&T model as simple as possible. In doing so, we leave process objects out of the T&T model, since process objects do not directly reflect the relationship between a registration and a product. Although we ignore the process objects, we keep the relationships between the registrations and information descriptions intact (see Fig. 4.5) The process object PO3.1 'Peeling machine 1' is not modelled but the relation with POM1 'peel scrape temperature' is still explicitly modelled between POA3 'assignation of machine' and POM1. The information instance II5, which was connected to PO5.1 'temperature sensor 1' in the process model, is now directly connected to POM1 'peel scrape temperature'.

4.3.2   Defining the relation types

In information space, only registrations are known. The trail of a product is, for the T&T system, nothing more than a set of registered idents. When products are assembled, the registered idents before and after the assembling differ, because the latter ident consist of one or more previous registered idents. To focus on these relations between registration points, we introduce between each two successive registration points a relation type. Be aware that relation types are related to processes but are not the same. Figure 4.6 shows an example in which three physical processes between two registration points are equal to one relation type in information space.

We have clustered the relation types into four groups as depicted in Fig. 4.7. The first group deals with the grouping of idents. These are reversible relations, since idents can be inserted into and separated from a group. The need for this type of operation is the change of granularity of an ident. When cans of tomatoes are packed into a box, the operations applied to the box do also apply to the cans in the box (e.g. change of location).

Group examples Assume that A, B, C and D are cans of peeled tomatoes. Let G be the box in which the cans are placed. The operation Create Group (CG) records that the cans A, B, C and D are indeed placed into box G. Operations on G are also relevant for A, B, C and D. When the cans C and D are taken from box G, this operation is represented by a separate from group (SFG). Operations on G are no longer applicable to cans C and D. If cans C and D are added again into box G, this should be denoted by a Insert in Group (IIG). A remove group (RG) describes the event of discarding box G.

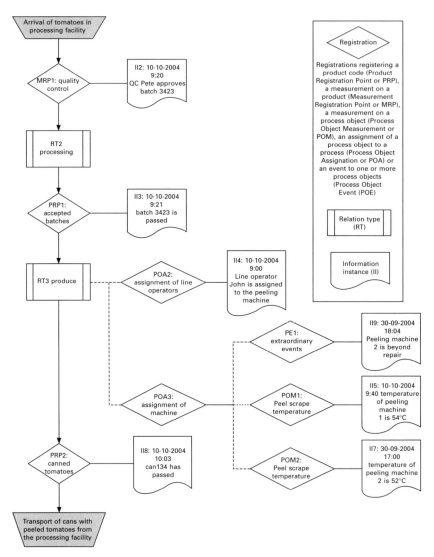

Fig. 4.5   Tracking and tracing model of the production process, shown in Fig. 4.3, of peeled tomatoes in a processing facility. The scope boundaries are indicated by grey trapeziums. The diamonds indicate a registration. Extraordinary events can be related to any process within the scope boundaries. This is why they are separately modelled.

The other three clusters are about non-reversible relationships. Transformations are about bringing idents together creating a new ident instantly. This type of operation is used when a new product emerges from other products, like lemonade from water and syrup.

Transform examples. A join represents the following event. Let A be a peeled tomato batch of farmer Johnson and B a peeled tomato batch of

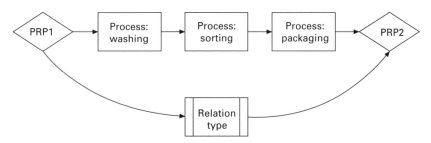

Fig. 4.6   The coherence between processes and relation types. Three physical processes are represented by one relation type in information space, because only two registration points are present.

farmer Peterson. Those batches are put together in a huge blender. Let C be the tomato juice that is extracted from the blender to a next process step.

Let A be the batch of fresh tomatoes, let B be the batch of peeled tomatoes and C be the peel residue. This is represented by a split relation type.

When a huge pile of tomatoes (A) is dispatched over small crates of tomatoes (B), it illustrates a produce situation.

Binding is about tapping an ident from an ident or adding an ident to an ident. We need a relation type which is capable of expressing a non-reversible addition to an ident, like adding salt to a pizza.

Bind examples. A typical Ident Creation From Ident (ICFI) situation is when a cow (A) gives birth to a calf (B). The cow-mother (A) still exists after this event.

When water (B) is added to a tomato batch (A), the identification of the tomato batch (A) remains the same, although an ident was added. An Insert In Ident (III) represents this situation.

Modality changes are about expressing changes to an ident without adding or extracting idents to/from it, e.g. heating a pizza in an oven. The relation type Processing represents this situation where A stands for the pizza.

In Fig. 4.5, decision point DP1 between the registration points MRP1 and PRP2 was translated into the relation type processing, because the outgoing idents are equal to the ingoing idents. Furthermore, the Processes P2 'Peeling tomatoes' and P3 'canning tomatoes' between the registration points PRP2 and PRP3 were translated into relation type 'PRODUCE', since the outgoing idents (can 134) are different from the ingoing idents (batch 3423).

4.3.3   Placing the process object related registrations
Registrations are taken from the process model that is attached to the process objects. These registrations are in the T&T model directly attached to the relation types to which the process objects are related. From the process

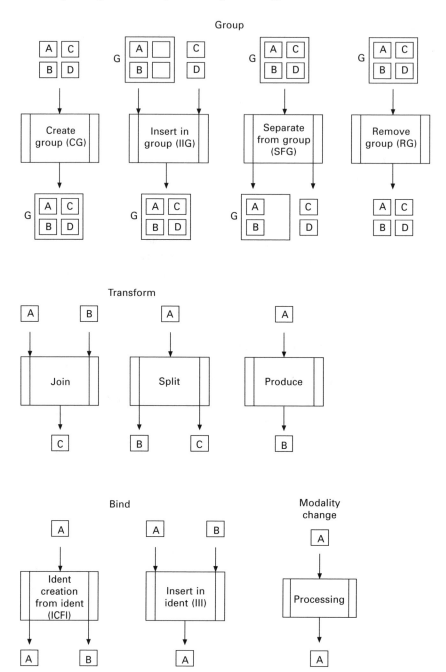

Fig. 4.7   Tracking and tracing relation types clustered into four groups of reversible and non-reversible clusters representing transforming (reversible), transforming (non-reversible), processing and binding.

model (Fig. 4.3), to process P3 peeling tomatoes and process P4 canning tomatoes, all attached POAs and POMs are taken (see Fig. 4.5). Those registrations are directly attached to relation type RT2. Note that the order of registrations remains the same as in the process model. This means that POA3 is before POM1 and POM2. As with POE3, a registration with the same identification is denoted only once.

4.4   Process and product issues in tracking and tracing modelling

In this section, we show that the T&T model as described in the previous section forms the basis structure for modelling T&T functionality in food supply chains. We now describe two important T&T issues and show how this can be incorporated into our model. The first issue is about ident mapping, addressing relevant aspects about the trail idents make in the T&T system when it is registered by a registration point. These aspects are an extension of the relation type. Is assigning a relation type enough to know the relation between ingoing and outgoing idents or do we need more information? The other issue is about the uniqueness of the identification of a product. This so-called Ident resolution is an indicator of the smallest traceable unit of products that can be recalled. Both issues bring in advanced features to the model. At the end of this section, a fragment is shown of a T&T model enhanced with these features.

4.4.1   Ident mapping in the tracking and tracing model

Processing a product takes time, this fact complicates traceability. A canning machine for example, takes ten minutes before the cans have travelled between the registration points before and after the machine. This makes it difficult to know which ingoing idents are related to which outgoing idents. For tracing purposes, it is essential to know the relationship between ingoing and outgoing idents of a relation type (e.g. knowing which tomato batch is inside a can of peeled tomatoes). The characteristics of products affect the relationship between ingoing and outgoing idents in a great way. We can define a spectrum of T&T behaviour where all products fit in. On one side of the spectrum there are the **discrete products**, that travel through (a part of) a chain one by one, like cans of tomatoes.

Both types of products are illustrated by the following examples followed by an explanation.

In a warehouse wrapped pallets are stored, each holding six boxes of tins with peeled tomatoes. These pallets have to be unpacked for small supermarket deliveries. A registration point named PRPin registers the full pallets at 9:00, 9:10, 9:20 and 9:30 (see Fig. 4.8). It takes some time to take the first box out of the pallet, since the wrap around the pallet has

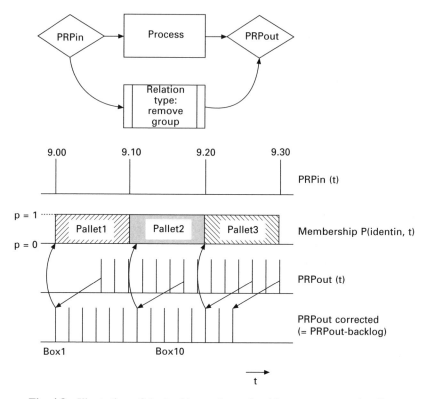

Fig. 4.8   Illustration of the backlog and membership concept around a discrete product.

to be removed. The unpacked boxes are registered by registration point PRPout. In order to know the relation between the ingoing pallets and outgoing boxes, we need to know the time between the registration of the pallet at PRPin and the registration of the first box at PRPout. This time is called the backlog. It is the correction factor necessary to find out what pallet was the origin of which box. The correction is made in the registration time of PRPout by applying the formula PRPout(t) corrected = PRPout(t-backlog). The backlog can be a constant time interval (e.g. 9 min) or a function depending on one or more relevant factors. The arrows in Fig. 4.8 indicate which pallet the outgoing box originated from.

On the other side of the spectrum are the **diffusive continuous products**, that travel through (a part of) a chain in a continuous stream and mix completely in an instant, like milk of two different origins poured into one tank. We illustrate this type of product with the following example shown in Fig. 4.9. In a milk factory, milk originating from different farmers is put into one tank. This is registered by PRPin. Milk packs are filled continuously. This process is registered by PRPout. Figure 4.9 shows the traceability of diffusive

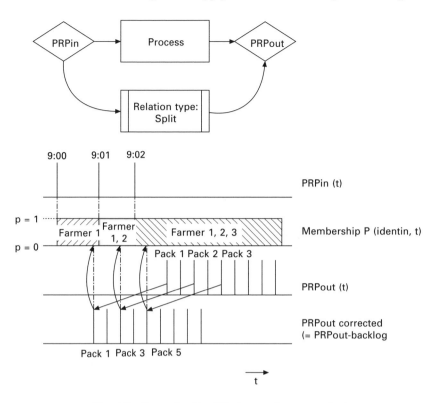

Fig. 4.9 Example of a diffusive continuous product.

continuous products. Taking into account that the backlog time is known, it is clear that Pack 1 originates from Farmer 1. Diffusive continuous products added together tend to mix instantly throughout the whole content. This implies that as soon as the milk of Farmer 2 was put in the tank, it was no longer possible to distinguish whether the milk inside the tank originated from Farmer 1 or Farmer 2. A similar story applies when the milk of Farmer 3 is put in.

Therefore we introduce the concept of membership. This is a function which represents the probability, on a scale of 0–1, that an ingoing ident is a member of all ingoing idents in time. So the membership function for Farmer 1 is 1 as long as the milk tank is not empty and decontaminated. The membership function of Farmer 2 is 1 from the moment that the milk was registered at PRPin until the tank is empty, and so on. Taking into account the backlog time, it is clear (see Fig. 4.9) that Pack 3 could be either from Farmer 1 or from Farmer 2. From Pack 5 on, the milk could be either from Farmer 1, Farmer 2 or Farmer 3.

Any product between those extremes are so-called **non-diffusive continuous**

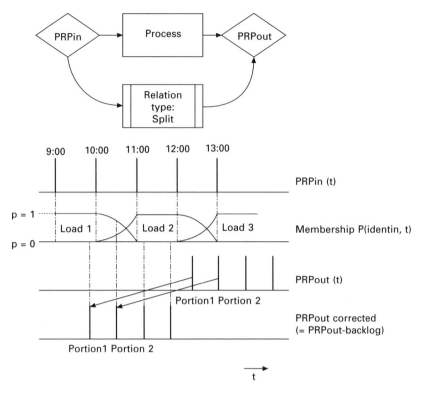

Fig. 4.10   Example of a non-diffusive continuous product.

products,[1] which travel through (a part of) a chain in a continuous stream but do not mix completely, like peanuts through a high-pressure tube. We illustrate this type of product with the example shown in Fig. 4.10. Different loads of the non-diffusive continuous product peanuts are put into a huge silo at different timeframes. This is registered by PRPin. From the silo, batches of peanuts are taken and registered by PRPout. Since the separation between two loads in that silo is not particularly clear, the membership function knows a so-called 'grey area'. In this case portion 2 (after the backlog correction) has a likelihood of about 60% that it originates from load 1 and 40% likelihood that portion 2 originates from load 2. In case of a need to recall load 2, a threshold of probability needs to be defined, in order to let portion 2 be recalled or not. If, for this example, the threshold is

---

[1] One could say that milk is actually a non-diffusive continuous product because it does not mix instantly in a huge tank. It is true that, looking closely enough, in fact no product is a diffusive continuous product. However, when a) the time steps are rather large, b) the batch size is small or c) the potential food hazard is high, we treat those products as diffusive continuous products. As soon as the membership function has probability values other than 0 or 1, a product is a non-diffusive continuous product.

lower than or equal to 40% likelihood, portion 2 needs to be recalled. If this threshold exceeds the 40%, portion 2 does not have to be recalled.

4.4.2 Ident resolution

In the previous section, we focused on features of relations between idents, now we pay attention to the features of the idents itself. Each product can contain either a unique identifier or a batch code, or carry no code at all. The magnitude of the physical products carrying the same identification is called the ident resolution. In the literature, there is also the term STUNT (Smallest Traceable Unit) (Vernède *et al.*, 2003). This resolution is important because it determines the absolute minimum amount of product that can be involved in a recall. Also it acts upon quality monitoring indicating whether there is knowledge of every individual product of every batch or just on the product in general.

Ident resolutions can be of three different types (see Fig. 4.11). The most detailed identification is the one where every product that passes a registration point has its unique identification. A less detailed identification is possible when the unique identification belongs to a product batch instead of the individual products, and the individual products all carry the physical batch code with them, e.g. all cans of batch 3423 have this batch code printed on top. The ident resolution here is equal to the batch size. Registrations and measurements to idents (PRPs and MRPs) are stored at a batch level, since the individual batch members are, from a T&T perspective, indistinguishable from each other. A critical measurement of one product of the batch can, therefore, result in the total recall of the whole batch, instead of that one faulty product. When products carry no physical identification at all, they can only be registered manually by an operator and described according to their contents. Thus, operator Pete, busy on process 'packaging', is registering a package containing strawberries by hand. The ident resolution has a near infinite size (i.e. the size of all the existing products in the world).

The ident resolution can be reduced by resizing batchsizes (each batch with its own unique identification). Thus, instead of putting 50 kg of fresh

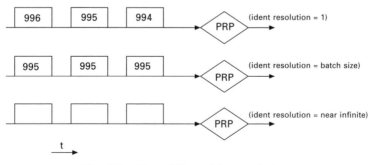

Fig. 4.11   Three different ident resolutions.

tomatoes into one crate (with a unique identification), 25 kg of tomatoes are put into a crate. The ident resolution for this particular ident is decreased by 50%.

4.4.3   The relevance of membership, backlog and ident resolution

If one tries to reduce the magnitude of a recall (like recalling 100 milk packs instead of 10 000) it is good to know that the focus should not be only on the identification, such as reducing the ident resolution by resizing the batch sizes. Suppose, for the sake of argument, that milk from a farmer delivered to the milk factory is not identified by the farmer but also by the farmer and the cow. This means that the ident resolution of the milk is relatively low. Assuming that the milk from all the farmers is still put into the same, not emptied and decontaminated, tank. This implies that despite the fact of a low ident resolution, all milk idents are part of the same recall. The membership functions reflect this example. When one is able to 'shorten' such a function it reduces the number of relations between outgoing and ingoing idents, which reduces the magnitude of involved idents in a recall. An example of shortening a membership function is to change the production process by adding a separate milk tank where only one ident is allowed to be put in. After the tank is empty, the process switches to the other tank while this tank is being cleaned and decontaminated.

When there is a membership function which is not steep on the flanks, e.g. with non-diffusive products, such as in the peanut silo example, the disadvantage is that so-called 'doubtful idents' are involved in recalls with a likelihood of less than 100% to be originated from more than one ingoing ident. When the flanks are steep, the time that 'doubtful idents' are related to ingoing idents is shorter, thus reducing the magnitude of a certain recall. In the peanut example, the 'grey areas' can be reduced by mounting a mechanism on top of the silo that flattens the peanut heap inside the silo before a new load of peanuts is put in. The grey area is now much smaller and therefore the flanks of the membership function are steeper.

In Fig. 4.12, we mapped the objectives to 'shorten' or 'steepen' the membership functions to the type of product. The mapping indicates where attention should be paid, i.e. in the identification (i.e. lowering the ident resolution by, e.g., reducing batch sizes) or in the process (e.g. extra tanks, reconstructing decontamination procedures or flatten mechanisms). Note that, after shortening the membership function of a continuous diffusive product, the membership function has the same shape as a discrete product. Therefore, if one wants to further shorten the membership function, the second improvement should be brought into the table as if it were a discrete product.

Since the backlog time is an estimator of the expected delay of a product during a process, it can be used to estimate the time it takes to travel through (a part of) the food supply chain by combining the backlog functions of all

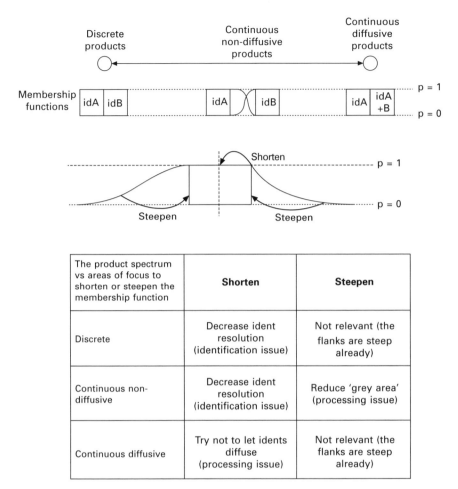

Fig. 4.12   Different types of product behaviour described as a spectrum; explanation of shortening and steepening.

relevant relation types. This can affect the recall procedure: for short chains a recall could be useless because all products involved are consumed already. For longer chains, the detected anomalies can be corrected before it even creates a food hazard for the consumer.

Now that we know that ident resolution, membership and backlog contain relevant information about the relations between ingoing and outgoing idents, we can add these concepts to the T&T model (see Fig. 4.13).

4.5   Future trends

Current generic T&T frameworks like FoodTrace (Foodtrace, 2004) and

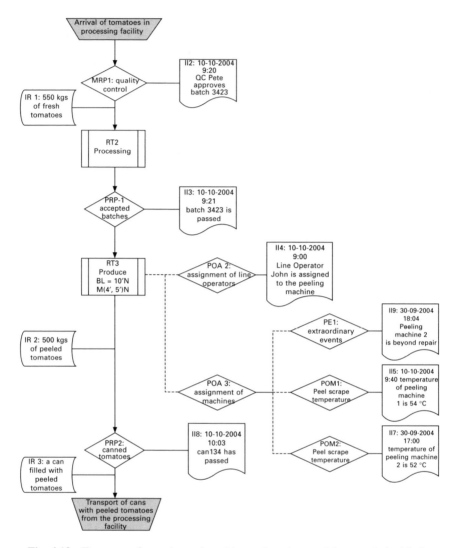

Fig. 4.13   Fragment of an enhanced tracking and tracing model, expanded with the concepts membership, backlog and ident resolution. This fragment is based on the tracking and tracing model in Fig. 4.5. Relation type 3 'Produce' describes the backlog as BL=10' N meaning that the backlog time is 10 min with a normal distribution. The membership function (M) is described here as the triple 4', 15', 4' N, meaning that the uprising flank is 4 min, the duration that the probability is equal to 1 is for 15 min succeeded with the down flank lasting for 4 min. All durations are expected to behave as a normal distribution (N). The ident resolution is reflected in the symbols IR1 through IR3.

CanTrace (CanTrace, 2004) focus on the mandatory duties of food supply chain participants, prescribed by legislation. Questions regarding how to design and implement the T&T system are left to the specific situation or project, or are left as an interesting subject for future research. The FoodPrint method focuses on how to design a T&T system, giving clear guidelines of relevant actions to undertake. The structured approach of creating a process model and T&T model are good examples of a) charting a sound representation of a food supply chain and b) creating a basis for thorough bottleneck analysis.

Legislation related to traceability in food supply chains tends to broaden its domain of interest. Apart from mandatory duties on traceability regarding the product itself, it also tends to legislate the traceability on the packaging of the product and to the related process objects (Vernède, *et al.*, 2003). The FoodPrint method and the process model take that broadened area of interest into account by modelling the processes, related process objects and registrations regarding process objects.

4.6 Conclusions

We are convinced that a systematic approach is essential to analyse, realise and improve T&T in food supply chains. Within such a systematic approach it is crucial to know what parts from the food supply chain are relevant to capture. During the development of the FoodPrint method, we identified those parts and put them into two perspectives; the food supply chain perspective, and the information perspective. Each perspective has its own structured approach and modelling techniques. The food supply chain perspective is expressed into the process model; the information perspective is reflected by the T&T model. Until now, we have not found any similar structured modelling approach in the literature.

4.7 References

CanTrace (2004), http://www.can-trace.org.
Chainfood (2004), http://www.chainfood.com.
Foodtrace (2004). Foodtrace Concerted Action Programme Generic Framework for Traceability Framework Considerations, FoodTrace http://www.cantrace.org/about/docs/1_multipart_xF8FF_2_EUFoodTrace%20FrameWork_details.pdf.
Vernède, R *et al.* (2003), 'Traceability in Food Processing Chains. State of the art and future developments', Agrotechnology & Food Innovations bv, The Netherlands.

5

Dealing with bottlenecks in traceability systems

N. Koenderink and L. Hulzebos, Wageningen University and Research Centre, The Netherlands

5.1 Introduction

Companies involved in the production or distribution of food products are often confronted with legislative requirements to install a tracking and tracing (T&T) system. The legislative purpose of such a T&T system is to monitor food safety and to find the culpable party in case of food risks. Some food chains, however, decided to use such a T&T system for more than just those defensive purposes. They used the opportunity created by the need to implement a T&T system to realise some of their strategic and tactical goals at the same time. We helped to design T&T systems in those food chains and consequently treated T&T as just one of the objectives of the system, while taking the other goals into account (see Chapter 2)

The perception that T&T systems can be used for other than defensive purposes, inspired us to develop the FoodPrint method. FoodPrint is a systematic method for creating purpose-oriented T&T systems. The method consists of five phases. In the first stage, an inventory is made of the operational goals of the food chain (see Chapter 2). As a second step, a model of the targeted food chain and the corresponding model of the T&T functionality are created (see Chapter 4). These models and the operational goals serve as a starting point for bottleneck analysis of the FoodPrint method, which aims to look at the current registration of product and process information with respect to the operational goals and to check whether the information system suits these goals. If the information system does not meet these goals, an analysis is made of the possible adjustments to the process itself, the information registration, the organisational structure and the technology used. This list of adjustments is used as input for the fourth phase of the FoodPrint method. In this step, the results of the bottleneck analysis are translated into a full system design. As a last stage, the T&T system itself is implemented. A general overview of the FoodPrint method can be found in Chapter 2.

In this chapter, we focus on the third step in the FoodPrint method: the bottleneck analysis, illustrated by a forest fruit quark case. This is a fictional case study in which two partners in the production chain of forest fruit quark are interested in sharing information via a T&T system. This case study is used in the remainder of this chapter to illustrate the different kinds of bottlenecks and the five steps of the bottleneck analysis. We continue with our view on future trends and we end with some conclusions.

5.2   Case study: forest fruit quark

To illustrate the FoodPrint bottleneck analysis, a simplified production chain of forest fruit quark is used. Forest fruit quark is a multi-ingredient food product composed of two main ingredients: forest fruit and quark. In this case study, we consider the production chain consisting of the fruit processing facility and the quark factory. The suppliers[1] of this chain are the raspberry, strawberry, blackberry and blueberry plantations, the dairy farmer, the producer of starter cultures, the producer of rennet, and the provider of packaging material. The vendor of the chain is the retailer. This chain is displayed in Fig. 5.1.

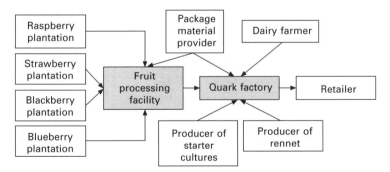

Fig. 5.1   Production chain for forest fruit quark.

---

[1] From the point of view of FoodPrint, suppliers are not part of the production chain under inspection (see Chapter 4). Suppliers do not take part in the information exchange and are therefore outside the scope of the T&T analysis. This implies that the production processes at the supplier's plant lie also outside the scope of the T&T focus. As a consequence, the product information that is delivered by the supplier is taken at face value. For the same reason, the vendors are also not considered part of the production chain.

5.2.1   The fruit processing facility

The production process in the fruit processing facility is displayed in Fig. 5.2. All four components of forest fruit – strawberries, blackberries, raspberries and blueberries – are transported to the forest fruit factory by the corresponding supplying plantations. The fruits arrive in crates of approximately 150 kg

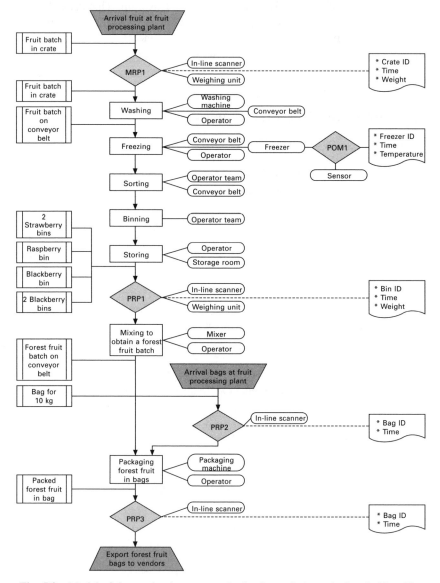

Fig. 5.2   Model of the production process in the forest fruit producing facility. The scope boundaries are indicated by dark grey trapeziums, the registration and measurement points (PRP, MRP, POM) by light grey diamonds.

each. Each crate is identified by a barcode. This code is scanned and the corresponding identification symbols, arrival time and weight of each crate are registered in the information system of the fruit processing facility.

The fruit from each crate is put on a conveyor belt. It is washed, frozen and sorted, and put in six identifiable bins, holding 25 kg each, which are stored in the storage room at a constant temperature. For the mixing of a forest fruit batch, 2 bins of strawberries, 1 bin of blackberries, 1 bin of raspberries and 2 bins of blueberries are retrieved from the storage room. The retrieval time, the bin code and the weight of the retrieved bins are registered in the information system of the fruit-processing factory. Thereafter, the fruits from the registered bins are mixed. A package material producer is the supplier of bags in which 20 kg of forest fruit can be packed. Each batch of bags has its own code. The batch code is scanned just before the bags are used in the process. The code and the registration time are stored in the information system. The forest fruit mix is packed in the registered bags, the filled bags are registered and subsequently transported to the quark factory, described in the next section.

5.2.2 The quark factory

The production process of quark in the quark factory is depicted in Fig. 5.3. The dairy farmer delivers milk in tanks to the quark factory. The arrival time and the tank code are registered in the information system of the quark factory. The dairy culture supplier and the rennet supplier both deliver their products packed in portions, each identifiable by their own barcode. On arrival, the portion codes, the weight of the portions and the arrival time are registered and stored in the information system. The quark production process starts with the pasteurisation of the milk. When the milk is pasteurised, a dairy culture is added. This dairy culture decreases the pH of the milk. When the pH is low enough, rennet is added to prevent clotting of the product. The mixture is tested regularly to determine whether quark formation is complete. Each time the operator stores the status of the mixture, the tank code and the measurement time in the information system. When the operator observes that the product is ready, the whey is removed. The tank code in which the quark is contained, its weight and the registration time are stored in the information system before the quark is transported to the forest fruit quark production line.

The final step in the process is the preparation of the forest fruit quark. The corresponding process is displayed in Fig. 5.4. The tank with quark and the required number of forest fruit bags arrive at the forest fruit quark production line. Their respective product codes and the arrival times are registered and stored in the information system of the quark factory. A supplier of cups and boxes delivers his or her goods in batches. Each batch of cups is registered in the information system, when they are taken from the storage room. Subsequently, when the forest fruit and the quark have been mixed in the

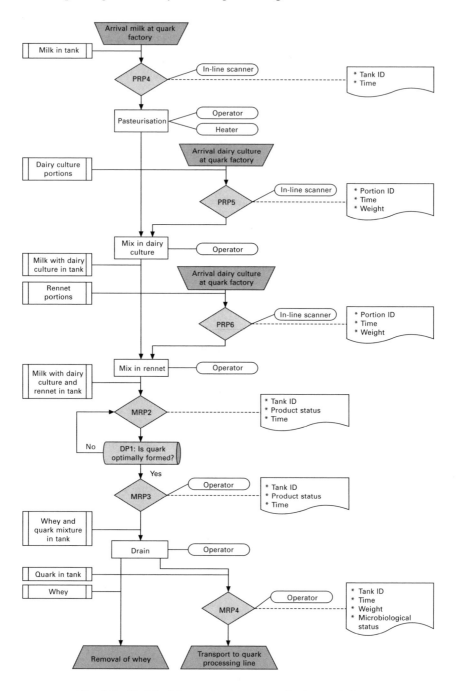

Fig. 5.3   Model of the production process in the quark factory.

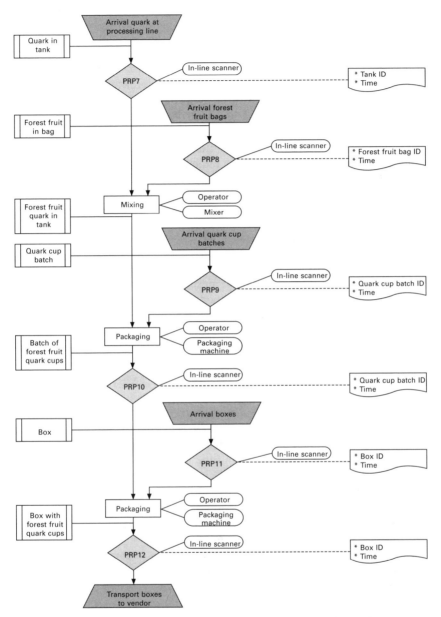

Fig. 5.4   Model of the production process of forest fruit quark.

correct ratio, the resulting forest fruit quark is packed in the cups. The cup codes and their filling time are registered in the information system. The cups are packed in boxes, such that each box contains 90 cups. The boxes are then prepared for transport to the retailer. When the boxes are packed, they are registered and the box code plus export time is stored in the information system.

5.2.3   Operational goals

The forest fruit producing facility and the quark factory that form the forest fruit quark chain are interested in tracing the ingredients on the granularity level of input batches. This means that, e.g. from a box of filled quark cups, the code of the original strawberry batch should be retrievable.

The chain partners have defined a second operational goal for the traceability system. They want to be able to find the source of any occurring microbiological contaminations. This means that, e.g. if there are Listeria monocytogenes traces found in a quark cup, the time of the registered contamination and the identification of the contaminated ingredient or process object needs to be retrievable.

5.3   FoodPrint terminology

To describe the method for bottleneck analysis as defined within the FoodPrint method, we use part of the terminology described in Chapters 2 and 4. In this section, we briefly recall the relevant concepts for bottleneck analysis.

The purpose of the FoodPrint method is to implement a T&T system. This is an information system that fulfils the requirements that are formulated in operational goals. When the model of the current food chain is made, there is usually some information system present in which data are stored. This system may or may not have already been used for T&T purposes; in either event it is not considered as a T&T system until the bottleneck analysis with respect to the formulated operational goals has been finished.

To be able to track and trace a product throughout the food chain, we need to make a distinction between the physical world, in which the product undergoes all kinds of operations, and the information world in which the corresponding information is stored. This information world is 'inhabited' by idents. Roughly speaking, idents are representations of the physical product. They can only come into existence in the information world, when the corresponding product in the physical world has been registered at a product registration point (PRP) or a measurement registration point (MRP).

The processes that a product undergoes are numerous. These processes are generally denoted in a process model, representing the product flow in the physical world. As a counterpart of the process model in the physical world, we recognise the T&T model in the information world. This T&T model contains the connections between the idents. It indicates which relation type took place between two registration points. All possible types of relations are described in Chapter 4. In this section, we mention only those relations that occur in the forest fruit quark case.

5.3.1   Transform: split

A split operation occurs when the input ident[2] and the output idents are different. In the physical world, this means that two or more new products are created from the input product. This happens, for example in our case study, at the forest fruit producing facility, where one crate of strawberries is transformed in a number of strawberry bins.

5.3.2   Binding: insert in ident (III)

An insert in ident operation occurs when there are more than one input idents and when one of these input idents is also the output ident. In the physical world, this means that one input product incorporates the other input products. An illustration of this operation can be found in our case example when dairy culture and rennet are added to the milk. In the information world, the input idents (Id_milk, Id_dairyculture, Id_rennet) are represented after the operation by the pre-existing ident Id_milk. This operation is irreversible; it is not possible to undo the mixing and reobtain the dairy culture and rennet separate from the milk.

5.3.3   Modality change: processing

A processing operation takes place when the identifier of the input ident is equal to that of the output ident, but when at least one of the properties of the product changes. This can be illustrated in our case by the example of the transformation from the product 'milk' to the product 'quark', simply by waiting. In that case, the nature of the product changes, but in the information world, the input and output ident are still the same.

5.3.4   Group: insert in group (IIG)

An insert in group operation occurs when a number of idents are grouped together and receive a shared ident. In a later stage, this shared ident could be removed and the original input idents can be processed separately again. An example of this can be found in the forest fruit quark at the point where the identifiable cups are put into a box.

5.3.5   Process information (POM & POE)

For some operational goals, it is necessary to store more information in the information system than just the product registrations. This can be done via process object measurements (POM). When the product status, e.g. the

---

[2] The input idents are the idents that are registered at the PRP or MRP directly before the operation, the output idents are the idents that are registered at the PRP or MRP directly after the operation.

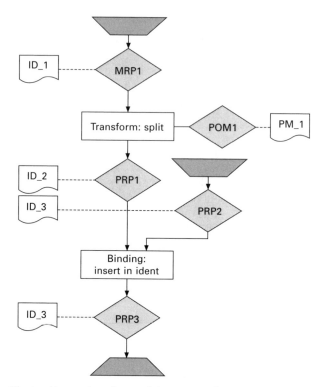

Fig. 5.5   The tracking and tracing model corresponding to the process model of the forest fruit producing facility.

weight, microbiological status or internal quality, is needed to fulfil an operational goal. this information can be entered in the information system via measurement registration points (MRPs). A last type of information that can be found in the information system is information about calamities or unexpected events, such as a broken machine or a power shortage. This information can be registered via process events (POEs).

5.3.6   The information models

From these definitions, we can create the process and information models of the current forest fruit quark chain. For the forest fruit quark case, this results in the T&T models as depicted in Fig. 5.5–5.7. In Fig. 5.5, we see the T&T model that corresponds to the process model of the forest fruit producing facility as displayed in Fig. 5.2. The process of washing, freezing, sorting and storing the forest fruits is modelled as a split operation, since the input crates are divided over multiple output bins. The mixing and subsequent packing of the forest fruits in bags is modelled as an insert in ident operation. In Fig. 5.6, we have depicted the T&T model that corresponds to the process model of the quark producing factory as displayed in Fig. 5.3. This model

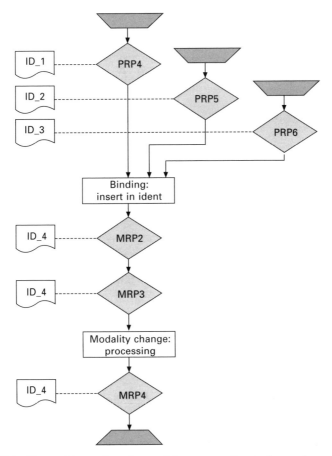

Fig. 5.6   The tracking and tracing model corresponding to the quark-producing factory.

contains an insert in ident operation, for the pasteurisation process followed by the addition of dairy culture and rennet and the waiting process. For the process of removing the whey, a processing operation is used. Figure 5.7 shows the T&T model that corresponds to the process model as displayed in Fig. 5.4. The mixing of the milk with the forest fruit and the pouring of the quark into cups is represented by an insert in ident operation. The packing of the cups in a box corresponds to an insert in group operation.

As mentioned before, in this chapter we focus on finding and solving bottlenecks in the traceability system. We define a bottleneck to be a point in the information stream at which it is impossible to find enough information about the product or production process to satisfy the information needs of the predefined business goals. The bottleneck analysis is a method to systematically track down the bottlenecks in the system model and solve them.

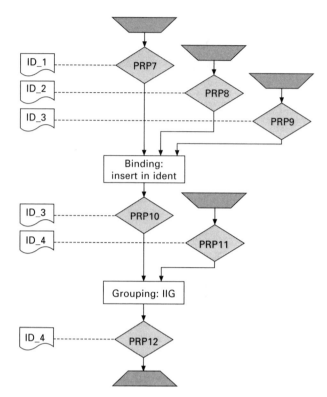

Fig. 5.7   The tracking and tracing model corresponding to the process of producing forest fruit quark.

5.4   Four types of bottleneck

Tracking and tracing solely takes place on the basis of information that has been stored in the information system. Traceability problems occur when the data in the information system do not represent the real world production processes in sufficient detail to meet the operational goals. When an operational goal is not met, a bottleneck needs to be solved. In that case, an adjustment in at least one of the factors: organisation, technology, information registration or process (see Chapter 2), has to be made, to ensure that, after the change, the stored information properly reflects the real world situation. We distinguish four different types of bottleneck.

5.4.1   Bottleneck due to insufficient product registration

The first type of bottleneck is caused by insufficient product registration. At a product registration point, but also at a measurement registration point, the product, its location and the registration time are recorded. The information thus stored in the information system should correspond to the real world history of the product. When the product is not sufficiently recorded during

Table 5.1   Some of the data stored at MRP1 and PRP1

Measurement registration points							
MRP1				PRP1			
Crate ID	Time	Weight	Type	Bin ID	Time	Weight	Type
Crate_id_1	9:00	149 kg	Strawberry	Bin_id_1	10:05	24 kg	Strawberry
Crate_id_2	9:05	150 kg	Blueberry	Bin_id_2	10:05	25 kg	Strawberry
Crate_id_3	9:10	152 kg	Raspberry	Bin_id_3	10:05	27 kg	Blueberry
Crate_id_4	9:15	148 kg	Blackberry	Bin_id_4	10:05	26 kg	Raspberry
Crate_id_5	9:20	151 kg	Blueberry	Bin_id_5	10:05	25 kg	Blackberry
Crate_id_6	9:25	150 kg	Blackberry	Bin_id_6	10:05	26 kg	Blackberry
Crate_id_7	9:30	150 kg	Strawberry	Bin_id_7	10:15	24 kg	Strawberry
Crate_id_8	9:35	147 kg	Raspberry	Bin_id_8	10:15	23 kg	Strawberry
Crate_id_9	9:40	149 kg	Strawberry	Bin_id_9	10:15	25 kg	Blueberry
Crate_id_10	9:45	151 kg	Blackberry	Bin_id_10	10:15	24 kg	Raspberry
				Bin_id_11	10:15	26 kg	Blackberry
				Bin_id_12	10:15	27 kg	Blackberry

the product's lifetime, the information world does not reflect the events in the real world well enough to support the reconstruction of the product trail as required by the operation goals.

In the forest fruit quark case, there are some examples of bottlenecks due to insufficient product registration. In Table 5.1, for example, part of the information database that corresponds to the washing, freezing, sorting and storing process of the individual forest fruit ingredients is displayed. According to the first operational goal, we have to check whether we can retrace the input crates corresponding to each bin. The information that is stored at measurement registration point PRP1, concerns the location and registration time of the bin codes Bin_id_1 to Bin_id_12. At MRP1, the location and registration time of the crate codes Crate_id_1 to Crate_id_10 are stored. This is all the information in the information world concerning the product history of the forest fruit ingredients in the washing, freezing, sorting and storage process. It is, therefore, impossible to tell whether Bin_id_1 originates from Crate_id_1 or from Crate_id_7. Thus, we have found a bottleneck for which the real world product history is not represented in sufficient detail by the data in the information world; the product registration is insufficient to enable the desired level of T&T. Possible solutions to this bottleneck are discussed in subsection 5.5.3. In this section, we restrict ourselves to indicating examples of the bottlenecks.

5.4.2   Bottleneck due to lacking process information

The second type of bottleneck is caused by insufficient registration of process information. When this kind of information is missing, the trail of the product may be correctly followed, but relevant process parameters are missing. These process parameters play an important role in food safety and food

Table 5.2   Some of the data stored at PRP7, PRP8, PRP9 and PRP10. This part of the process represents the mixing of the forest fruit bags (FFB_id_#) with the quark (MT_id_#) and the pouring of the resulting forest fruit quark into the cups (QC_id_#)

Measurement registration points							
PRP10		PRP9		PRP8		PRP7	
ID	Time	ID	Time	ID	Time	ID	Time
QC_id_1	16:48,00″	QC_id_1	16:46,00″	FFB_id_1	16:30	MT_id_1	16:29
QC_id_2	16:48,20″	QC_id_2	16:46,20″				
QC_id_3	16:48,40″	QC_id_3	16:46,40″				
QC_id_4	16:49,00″	QC_id_4	16:47,00″				
...	...	...	...				
QC_id_90	16:57,00″	QC_id_90	16:55,00″				

quality. An example of such a bottleneck is the lack of temperature information for decay sensitive products. It is, for example, not good enough to know that a product has been in cold storage; we also need to know the temperature of the cold storage.

In the forest fruit quark case, a bottleneck of the second type is encountered when we check the information system to find the source of microbiological contaminations. When we trace upstream, the first production step where pollution may occur is at the point where the forest fruit is mixed with the quark and the resulting forest fruit quark is packed into the cups. The corresponding relevant part in the information database is described in Table 5.2. For this production step, various operators that work at the production line, the machines that are used for the mixing and packing and the in-line scanners are involved. When one of the machines used is contaminated, it may be the undetected source of contamination of the forest fruit quark.

5.4.3   Bottleneck due to insufficient product measurement

The third type of bottleneck is caused by insufficient product measurements. When this type of bottleneck occurs, the trail of the product may be correctly followed and the process information may also be correctly registered, but information regarding the status of the product is missing. Such a bottleneck occurs for example when the microbiological status or the initial product quality is not measured. Then, even though the process information is correctly stored, the product may not fulfil the food safety or food quality standards of the production chain.

We encounter this situation in the forest fruit quark case at the production step that is described in the previous example. When either the forest fruit or the quark already contains a microbiological contamination, and when there is no check of this product status, the information system does not contain sufficient information to pinpoint the origin of the microbiological contamination at this production step. It is impossible to ascertain whether

Table 5.3 Some of the data stored at PRP10, PRP11 and PRP12. This part of the process deals with the packaging of 90 cups in each box.

Measurement registration points					
PRP12		PRP11		PRP10	
ID	Time	ID	Time	ID	Time
Box_id_1	17:00	Box_id_1	16:55	QuarkCup_id_1	16:48,01″
Box_id_2	17:15	Box_id_2	17:10	…	…
Box_id_3	17:30	Box_id_3	17:25	QuarkCup _id_90	16:57,00″
Box_id_4	17:45	Box_id_4	17:40	QuarkCup _id_91	17:03,01″
Box_id_5	18:00	Box_id_5	17:55	…	…
Box_id_6	18:15	Box_id_6	18:10	QuarkCup _id_180	17:12,00″

the contamination has been introduced in this process step, or whether it was already present in the input ingredients.

5.4.4 Bottleneck due to insufficient process event registration

The fourth type of bottleneck is caused by insufficient process event registration. In these cases, the information world may seem to represent the product history accurately, but it actually misses out on relevant information on process events. An example of such a bottleneck can be found in the forest fruit quark case at the last step in the production process. The end product of the forest fruit quark chain is a box filled with cups. In the information world, each such box is represented by an ident Box_id_1 to Box_id_6. For each box, the box and the registering time at product registration point PRP12 is stored in the information system. In Table 5.3, part of the information database can be found. We see that the input idents of the cups QuarkCup_id_1 to QuarkCup_id_90 and the input ident of the box Box_id_1 belong together. The same holds for the input idents QuarkCup_id_91 to QuarkCup_id_180 and Box_id_2. Therefore, it seems that there is no bottleneck at this registration point.

This conclusion, however, is only correct under the assumption that no unexpected events take place, e.g. that all boxes are filled properly or that the machine does not break down. Since there is no way to register the occurrence and type of calamities in the current information system, it is impossible to register these problems when they occur. Therefore, in the case of an unforeseen event, the real world process will not be correctly represented in the information world. This results in a bottleneck of the fourth type.

5.5 Analysing and resolving bottlenecks

Within the FoodPrint method, bottleneck analysis plays an important part. Only by performing this analysis, the functioning of the current information system can be tested and problems can be solved. Bottleneck analysis is an iterative process, consisting of five steps.

5.5.1   Ordering the operational goals

In order to perform a good bottleneck analysis, the goals of the traceability system must be specified in sufficient detail. Therefore, the first step in the bottleneck analysis consists of ordering the operation goals with respect to their relevance for the chain partners. By ordering the operational goals, we make sure that the bottlenecks for the most important objective are found and solved first.

For the forest fruit case, this means that the two specified operational goals for the traceability system need to be assessed with respect to their relevance. The chain partners indicate that the order of importance is as follows:

1. The forest fruit quark needs to be traceable on the granularity level of input batches.
2. The microbiological contaminations need to be traceable on the granularity level of bins (in the forest fruit producing facility), tanks (at the quark production line) and cups (at the forest fruit quark line).

5.5.2   Finding the first bottleneck

When the operational goals have been ordered, the first bottleneck with respect to the most important goal should be found. The tracing of the bottleneck can be done both upstream and downstream. Usually, the type of operational goal gives a logical direction.

In our example, the first operational goal states that from the final product of the food chain, i.e. the forest fruit quark, the input batches need to be traceable. When we trace bottlenecks for this goal, it makes sense to start with the end product and to trace the product history upstream. This is also true for the second operational goal. If, however, the partners would have defined a third operational goal that asks for a reduction of a recall, it would make more sense to follow the production process downstream.

5.5.3   Finding possible solutions for the bottleneck

In the third step of the bottleneck analysis, possible solutions to the bottleneck are formulated. A bottleneck of the first type, a lack of product registration, can be solved by adding a registration point in the process, such that the product registration is sufficient to fulfil the current operational goal. Another possibility is to change the process in such a way that the existing registration points record enough information such that the bottleneck does not exist anymore. A change in these two factors is often the most obvious solution for a bottleneck of the first type, although a solution in the identification technology used or the organisational aspects of the process might also be taken into consideration.

In the forest fruit quark case, we encounter a bottleneck of the first type, i.e. caused by insufficient product registration. To solve this bottleneck, we note that in the information world it is unclear whether, e.g., the strawberries corresponding to ident Bin_id_1 originate from the crate represented by Crate_id_1, Crate_id_7 or Crate_id_9. The output idents are not unambiguously linked to the input idents. This bottleneck is caused by two issues. The first is the random order at which the forest fruit components are taken out of the storage room. The second is the difference in batch size between the input crates of 150 kg and the output bins of 25 kg in combination with the long time interval.

There are several ways to solve this traceability problem. One solution is to make an adjustment in the process by adding an additional registration point MRP1a before the storing of the products in the storage room. In that case, the 'first in first out' character of the washing, freezing, sorting and binning processes and the overall backlog time of the combined process is a good indicator for the coupling between output idents and input idents. For this process, the backlog time is equal to 25 min. The forest fruit ingredients are non-diffusive continuous products (see Chapter 4). To have a reliable indication of the probability that the product in a certain output bin originates from a certain input crate, the membership function needs to be determined. Once this is done, the coupling between input and output idents is as reliable as possible in this process.

The additional registration point MRP1a does not only enable us to determine the coupling between input and output idents. It also makes sure that the input bins are scanned and identified before they enter the storage room, which ensures that despite the random order in which the bins leave the storage

Table 5.4 Lay-out of the information system when an additional registration point is added before the bins are stored in the storage room. Thus, the bins corresponding to the idents that have been registered at MRP2 can be retraced to the original crates that have been registered at MRP1. Note that we have only displayed part of the information, the weight of the crates and bins is left out in this table

Measurement registration points							
MRP1			MRP1a			PRP1	
Crate ID	Time	Type	Bin ID	Time	Type	Bin ID	Time
Crate_id_1	9:00	Strawberry	Bin_id_1	9:25	Strawberry	Bin_id_4	10:05
Crate_id_2	9:05	Blueberry	...	...	...	Bin_id_2	10:05
Crate_id_3	9:10	Raspberry	Bin_id_6	9:29	Strawberry	Bin_id_9	10:05
Crate_id_4	9:15	Blackberry	Bin_id_7	9:30	Blueberry	Bin_id_16	10:05
Crate_id_5	9:20	Strawberry	...	...	...	Bin_id_21	10:05
Crate_id_6	9:25	Blueberry	Bin_id_12	9:34	Blueberry	Bin_id_11	10:05
Crate_id_7	9:30	Raspberry	Bin_id_13	9:35	Raspberry	Bin_id_1	10:15
Crate_id_8	9:35	Blackberry	...	...	...	Bin_id_5	10:15
			Bin_id_18	9:39	Raspberry	Bin_id_10	10:15
			Bin_id_19	9:40	Blackberry	Bin_id_7	10:15
			...	...	...	Bin_id_17	10:15
			Bin_id_24	9:44	Blackberry	Bin_id_24	10:15

room, the coupling between input bins and output bins is unambiguous. A relevant part of the corresponding information system can be found in Table 5.4.

Another way to solve this bottleneck is to change the process in such a way that the fruits are delivered in bins or repacked in bins before they enter the washing, freezing, sorting, binning and storing process and that these bins are registered before this process. In that case, the corresponding table looks almost the same as Table 5.4, with the only difference being that MRP1a is located before the products enter the washing, freezing and sorting process. In case the fruits are delivered in bins, the table corresponding to MRP1 is left out of the information system completely.

A third way to change the process is to change the amount of goods that are processed at the same time. We can for example make sure that the fruits are processed in crates instead of in bins. This lifts the level of detail to a larger scale, but is a good solution for the problem caused by the random order of products entering and leaving the storage room, since in this scenario the crates are identifiable.

Finally, we could also think of a solution in which the crate identification is transported with the product on the conveyer belt during the processing and a sequence number is automatically added to the bin code to identify which crate has delivered its contents to the bin, before the bins are put in the storage room.

When the second step of the bottleneck analysis has found a bottleneck of the second type, i.e. a bottleneck caused by insufficient registration of process information, a change in the information that is recorded needs to be made. The extra information that needs to be stored in the information system can be obtained by a change in the organisational aspects of the production process, or by a change in the technologies used.

In the forest fruit quark case, we encounter a bottleneck of the second type, i.e. caused by insufficient registration of process information. This example describes the inability of the information system to find possible microbiological contaminations because the microbiological status of the mixer is not well registered. To make sure that this bottleneck is solved, the

Table 5.5 Cleaning record of the mixer, corresponding to the solution where an operator checks the microbiological status of the mixer

Measurement registration point		
POM2		
Time	Operator	Microbial status
9:00	Pete	Microbial values below threshold
11:00	Pete	Microbial values below threshold
13:00	John	Microbial values below threshold
15:00	John	Extra cleaning is necessary
15:15	John	Microbial values below threshold

organisational aspect of the process can be changed in such a way that the person who determines whether the mixer needs cleaning enters the microbial status of the mixer into the information system. The corresponding data in the information system may look like those in Table 5.5.

A second solution is to add a measurement device that continuously checks the mixer for possible contaminations. When this device detects that acceptable levels of microbiological activity are exceeded, the mixer is automatically stopped and the machine has to be cleaned.

A bottleneck of the third type, a lack of product measurement, is typically solved by adding measurement registration points. In subsection 5.4.3, we mentioned a bottleneck in the second operational goal (to find the source of microbiological contamination). We noted that, when the input forest fruit or the input quark is already contaminated and this is not measured and registered, the source of the microbiological contamination cannot be traced. To solve this bottleneck, we can add two measurement registration points before the forest fruit and the quark are mixed, that determine the microbiological status of the forest fruit and the quark respectively.

To formulate solutions for a bottleneck of the fourth type, i.e. a bottleneck caused by insufficient registration of process events, we need to store additional information in the information system. In order to know which information needs to be added, we can make an inventory of the process events that may occur. With this information, we can establish which process object could be sensitive to process events. On the basis of this information, solutions for bottlenecks of the fourth type can be presented. This is usually done in the form of an adjustment in the organisational aspects, making sure that an operator is responsible for entering the data in the system and indicating in which way this should happen.

In the forest fruit quark case, we encounter a bottleneck of the fourth type when the cups are packed into boxes. This situation has been described in subsection 5.4.4. To solve the problem of not having the possibility of storing information concerning calamities, we can simply create the possibility of entering process event information into the system. Therefore, we make an operator responsible for recording data concerning process events for this process. In case of a calamity, the operator must enter the corresponding process event information into the system via a user interface. He or she needs to record the start and end time of the process event, the process object that was affected and the nature of the event. In the case of the malfunctioning package machine, the operator would have to enter a description such as 'The package machine ran out of duck tape, which caused the bottom of the boxes to open and the quark cups to fall out'.

5.5.4 Choosing the best solution

In most cases, there will be more than one solution for the found bottleneck. In order to decide on the best solution, a cost–benefit analysis has to be done, in close communication with the chain partners. On the basis of this analysis, the best solution can be chosen and the process and information analysis will be changed accordingly. For each solution, we need to make sure that the newly chosen solution does not introduce bottlenecks for the already covered operational goals.

5.5.5 Finding further bottlenecks

The steps that find bottlenecks, formulate solutions and choose a solution are to be repeated until all bottlenecks for each operational goal have been solved. When the whole process is finished, the result is a number of proposed changes that will transform the existing food chain that does not meet the formulated operational goals into one that does meet these requirements.

5.6 Future trends

We expect that the focus of T&T will change in the near future. At present, T&T is often seen as a goal on its own, imposed by external parties to ensure food safety and pinpoint the liable party for food safety problems. We expect that companies and networks of companies will start using their T&T systems as a means to reach other goals. When not only the location of the product and the occurrence of process events, but also the process parameters and the product measurements are registered, a company or production chain can use this information, e.g. to aim for an improvement of product quality. When the information system is used for such operational goals, it will be important to be able to check, for each operational goal, whether the existing information system contains sufficient information. The bottleneck analysis as described in this chapter is a good method to systematically check and improve the functionality of the information system.

5.7 Conclusions

The bottleneck analysis described is a systematic method to track down bottlenecks of all four types in the food chain under consideration. By integrating bottleneck analysis with operational goals the system models, we have created a strong method that enables food chains to turn the (legislative) pressure to implement a T&T system into an opportunity to realise various other operational objectives and thus to strengthen their competitive position.

6

Including process information in traceability

M. Klafft, Institute of Information Systems, HU Berlin, Germany,
J. Huen, C. Kuhn, E. Huen and S. Wößner, Fraunhofer Institut
für Produktionstechnik und Automatisierung, Germany

6.1 Introduction: benefits for the industry and the consumer

The issues of quality and safety are of overwhelming importance for the food industry. Quality aspects, such as the organoleptic properties of the food (taste, smell, outer appearance) and its nutritional properties are decisive selling arguments and finally determine a product's economic success. Food safety is of even greater importance, as such safety problems endanger the health of the consumer and can ruin entire companies due to the negative impact on the manufacturer's image. However, the emergence of new diseases like TSE (transmissible spongiform encephalopathies) and several food scandals (e.g. dioxin or nitrofen contamination of chicken) have demonstrated that – despite intense quality assurance efforts by the industry – food safety problems still occur. In order to be able to recall all affected products in case of an emergency, the European Union adopted regulation 178/2002/EC (EU, 2002). This regulation requires each food producer and distributor to be able to provide the information on when and from whom they received a product for further processing and when and to whom they delivered a product (Waldner, 2004). Key industrial players such as Unilever went even further and implemented holistic traceability concepts over the whole supply chain (Kuhn and Huen, 2004), thus being able to identify immediately who was involved in the production of a defective batch.

Traceability, however, only deals with the whereabouts of the food products, which facilitates a recall, but does not provide any information or hints about the cause of the problem responsible for the recall. Thus, the recall will be unfocused and uneconomic. Being able to retrieve additional information on defective products, such as processing data, could reduce this problem, offering the following benefits:

- The cause and thus the origin of the defect can be identified more quickly.
- The number of products affected by a possible recall can thus be limited to the necessary minimum. Today, a recall can strike the entire supply chain, and even some related supply chains, blocking deliveries from many companies for days, until the products in question have been examined and released by the food authorities. However, if process information is included in traceability, problems related to inappropriate processing can be identified quickly, thus limiting the recall to the affected batches.
- Furthermore, a faster and more focused response reduces the risk that defective products reach the consumer. This limits both the health risks of the consumers and the negative impact that such health problems have on corporate image.

It is therefore beneficial to include process and product data in traceability. A general model of such a traceability system is shown in Fig. 6.1.

6.2   Using process information to improve quality

Traditionally, product and process parameters have been defined by considering each parameter on its own and then deciding on acceptable thresholds guaranteeing product quality and safety. The fact that several quality-relevant parameters influence each other, was only intuitively taken into account. More recently, however, the idea that it is actually a combination of various product and process parameters that determines the quality of a product became more popular and resulted in the development of so-called hurdle technologies. Hurdle technologies aim at preserving the product by identifying acceptable combinations of food parameters. Such hurdles may include temperature, pH, water activity, redox potential, modified atmosphere and preservatives (Leistner and Gould, 2002). The difficult questions are, however, which of these combinations can be considered safe, and how do the food processing steps influence hurdle-related parameters. Very often the interactions of the various factors are not fully known and predictable, which makes it necessary to run experiments before releasing a product on the market – and this is where including process information in traceability comes into play.

Product traceability and the acquisition of corresponding process data makes it possible to manufacture products in an industrial production environment using varying process parameter combinations and still know afterwards how each product was treated. The resulting data set can be used as an input for statistical methods for parameter specification and, at a later stage, food quality prediction. This means that it will be possible during the production process to predict if a product is likely to become a reject or not.

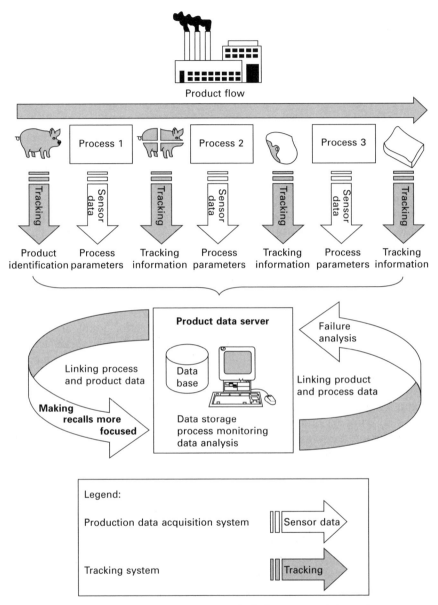

Fig. 6.1   A food traceability system including process information.

At a later stage, additional data gained during full-scale production can be used to constantly check and improve the models for data analysis.

These advantages mean that, including process data in traceability makes recalls more focused, can improve the food safety, delivers the information for flexible process parameter specification and facilitates food quality prediction during production.

The following sections deal with technologies and methods which are important for the implementation of systems including process data in traceability. The range of topics discussed includes sensors for data acquisition, product identification techniques, data handling and statistical methods for data analysis.

6.3   Methods for collecting and storing information

6.3.1   A general model linking product and process information

Before deciding to implement traceability systems including process information, a thorough evaluation should be made in order to determine whether such a system is useful for a specific application. Such an evaluation could proceed as shown in Fig. 6.2. The first step is to carry out a hazard analysis and to identify critical control points, following the (mandatory) 'Hazard analysis and critical control points' (HACCP) method. However, non-hazardous quality aspects should be considered as well. This leads to the identification of additional 'ordinary' control points, defined as points where a loss of control leads to economic or quality-related damage without endangering the consumer (for more details on control points and critical control points, see Humber, 1993). As a result of this process, parameters for food quality and safety, which will have to be monitored during production, are identified. The company has to decide which of these parameters are so critical that they require 100% monitoring and, if online control is the preferred solution, appropriate sensors have to be identified and then integrated into manufacturing execution systems and documentation systems. Based on the results of the analyses carried out so far and the technologies chosen for process monitoring, a final decision is made on whether to include process information in traceability. Questions affecting this decision are:

- Is including process information in traceability desirable for legal reasons (product liability, regulations such as EU (1985) combined with EU (1999), product-specific traceability regulations...)?
- Does the company intend to use a hurdle-technology approach, and does this require an experimental phase to identify acceptable parameter combinations?
- Is the production process and the influence of product and process parameters on food quality well understood?
- Does the company gain a competitive advantage by including process information in traceability?

If, finally, the decision is taken to implement traceability concepts including process information, there are two ways to implement such holistic systems at the production line: either by focusing on batch-related data or by including product-individual data.

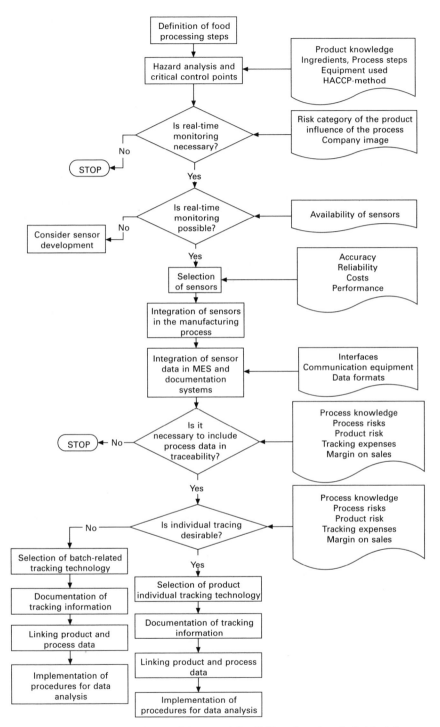

Fig. 6.2 How to determine usefulness and feasibility of product-individual data acquisition.

Today's quality assurance systems often use the batch-related approach, only storing information on samples taken from the batch, which have been analysed in a laboratory. This approach, however, does not reveal the full potential of including process data in traceability because analysing samples does not make it possible to identify very local or temporary problems. Monitoring overall process data of the batch does not guarantee that a specific product was treated properly (e.g. if it was temporarily stuck inside a machine).

These problems do not occur if each product and its corresponding processing data can be traced individually. Furthermore, product-individual traceability provides significant advantages during the ramp-up phase of a food product, especially during process and product parameter specification, making it significantly easier to analyse the influence of process parameter variations on product quality. On the other hand, product-individual traceability is only feasible if the product flow consists of insular products and it usually requires greater technological means and is thus more expensive.

Taking the above mentioned aspects into consideration, it is up to the food producer to decide which concept (or which combination of the two concepts) is most suitable for the specific production line in question. Then, methods and tools for documenting tracking information will be chosen and a link between the product and corresponding process data will be established. The last step to complete the system is the introduction of procedures for data analysis during the testing and ramp-up phase as well as for quality prediction during full-scale production.

6.3.2   Determining the suitable degree of process information

As discussed above, a very important point to be decided during the evaluation and implementation process is which information to include in the traceability system. This decision is mainly based on the analysis of the product-inherent risks and the risks inherent in the process. The following method, combining elements of the HACCP method with additional input from the 'Failure Mode and Effect Analysis' (FMEA) method, could be used to evaluate these risks:

- Identification of process steps (already included in the flow chart above).
- Identification of possible failures.
- Appraisal of possible damage caused by such a failure (scenario analysis, including safety aspects as well as quality aspects such as the effects of undesirable organoleptic and/or nutritional product properties).
- Identification of possible reasons for the failure.
- For each of the failures identified above, the following aspects have to be evaluated by awarding points between 1 and 10 (adapted from Ford Motor Company, 1996):
    - Severity of effect (1 point: no effect, 10 points: hazardous without warning).

- ▪ Likelihood of occurrence (1 point: remote, 10 points: very high).
- ▪ Likelihood of detection (1 point: almost certain, 10 points: very uncertain).
- A risk priority number is then calculated for each failure by multiplying the three point values:

Risk priority number = points(frequency) * points(damage) * points(detection)

Thus, the risk priority number may vary between 1 and 1000. In practice, many companies define a fixed threshold (e.g. 200) and consider all failures scoring above as unacceptable, though such limits are somewhat arbitrary (as the evaluation includes many subjective criteria).

- A first step is then to look for and to implement hazard avoidance measures, thus systematically reducing the risk priority numbers. This may include the definition of hurdles inhibiting, e.g., microbial growth.
- If significant risks remain, online sensors for automated 100% quality monitoring should be used, in order to control these hazards.
- If the risks are related to several hurdles influencing each other, and if these interactions are not well understood, using product-individual traceability and included process information would be a good solution to analyse and subsequently control these factors.

In the following sections, the product-individual approach will be discussed in detail, because it provides optimal conditions for data analysis and quality prediction. Various technologies needed for the implementation of such an approach will be presented, starting with sensors for online data acquisition, as they are the basis for all product-individual data analyses.

6.3.3   Sensors for process data acquisition

Sensor research has made tremendous progress during the last 20 years, especially as the enormous increase in computational power made it possible to analyse vast amounts of data in real-time. In addition to this development, existing sensor technologies were constantly improved and refined, so that both accuracy and reproducibility of the measurements increased significantly. Table 6.1 gives examples of already existing online sensor technologies, their advantages and limitations as well as possible applications in the food industry (adapted from Food Manufacturing Coalition for Innovation and Technology Transfer, 1997, if not indicated otherwise).

As can be seen, several product and process parameters can be monitored online, thus providing a solid basis for product-individual data acquisition and quality control concepts using individual traceability. One very tricky task, however, remains to be solved: the coupling of product and process data with the end product. Therefore, appropriate product identification techniques are needed.

Table 6.1 Online sensors for non-destructive real-time monitoring of food products (Source: adapted from Food Manufacturing Coalition for Innovation and Technology Transfer, 1997)

Technology	Application	Advantages	Limitations
X-Ray	Detection of foreign objects	Contact-free, can detect both metal and some non-metal objects (e.g. bones, plastics)	Requires strict safety measures (hazardous to personnel)
Microwave	Measurement of density and moisture content	Measures bulk properties of the material and not just the surface (Kent, 2001)	Expensive, implementation requires a lot of expert knowledge and careful design (Kent, 2001)
Pneumatic Probes with laser-based sensors	Measurement of firmness (modulus)	Contact-free, fast	Although the method is contact free, it generates an air puff which may pose a contamination risk
Nuclear Magnetic Resonance (NMR)	Determination of soluble solids, sugar content, solid fat content and ripeness	Fast	Expensive
RF impedance measurement	Determination of moisture content (for an example, see Schmilovitch *et al.*, 1996)		Not contact free, may need to contact every individual product, thus slowing down the analysis and posing a cross-contamination threat
Acoustically induced mechanical resonance	Measurement of firmness (modulus)	Analyses the entire product, e.g. fruit	Needs further development, so far only tested in a laboratory environment (Kilcast, 2001)
Videography	Determination of colour and size	Contact-free, fast	Cannot penetrate surfaces
Rheological sensors	Determination of texture, flavour release, stability,	Well developed, numerous technological alternatives	Many rheological methods require contact to the product

Technique	Application	Advantages	Disadvantages
	appearance (Roberts, 2001) as well as viscosity. Monitoring of dough development.	available	
Infrared remote thermometry (Ridley, 2001)	Temperature measurement	Measures temperature over a whole surface, not only at a specific point	Expensive, unable to measure core temperatures, requires free 'line of sight' to the product
Near-infrared-absorption (Benson and Millard, 2001)	Compositional analysis (e.g. moisture, fat, protein and sugar content, lactose content (Tsenkova et al., 1999), ergosterol (Farkas, 2003, Dowell et al., 1999), deoxynivalenol (Dowell et al., 1999), thickness measurements (e.g. of chocolate coatings on refiner rollers)	Fast, contact-free, independent of many varying product parameters (e.g. electrical properties)	Needs shielding from ambient light, limited penetration of the product. Not so suitable for shiny surfaces

6.3.4 Product identification techniques

The most common product identification techniques used today are barcodes and radio frequency identification tags – techniques which are discussed in detail elsewhere in this book and therefore will not be the focus of this section. One disadvantage of barcode and RFID systems is that an information carrier (e.g. a barcode label or an RFID tag) has to be attached to the product. This is easy for packaged products but difficult if the food product is still being processed. For non-pumpable products, one solution is to attach labels or tags to a product carrier such as a box or a tray. Such solutions are widespread in the meat industry. One example is the tracing of pig carcasses using a transponder in the hook on which the carcass is hanging (Harris, 2004). More sophisticated systems use image processing to monitor product pieces on conveyors or even during manual operations. The Fraunhofer Institute for Manufacturing Engineering and Automation (IPA) developed such a traceability concept for the fine cutting in the meat industry (Huen, *et al.*, 2003). The concept, named CarnTrack, consists of the following elements:

- A number of segregated work places (cells).
- A conveyor belt transporting pieces of meat between those work places.
- A loading unit which feeds the conveyor with pieces of meat of known identity.
- An image processing system which tracks the movement of meat pieces on the conveyor and detects if the pieces are taken from a conveyor into a cell or vice versa.

The layout of the CarnTrack concept is shown in Fig. 6.3.

The loading unit places the original pieces of meat on the conveyor at a defined position and simultaneously transmits identity data to the CarnTrack system. During transportation on the conveyor, the movement of the meat on the belt can either be monitored with the help of cameras or – if precautions

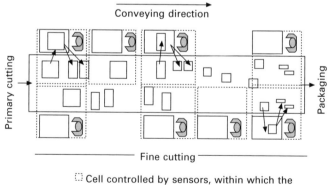

Cell controlled by sensors, within which the exchange of meat pieces between the conveyor and the individual working places is registered.

Fig. 6.3 CarnTrack – the principle (Source: translated from Huen and Klafft, 2004).

against slipping have been taken – be calculated by measuring the conveyor speed. Thus, the system always knows when which piece of meat reaches a cell. At each cell, another camera detects butcher interactions, i.e. taking pieces from the conveyor to the working station and vice versa. If each butcher only works on one piece of meat at a time, piecewise meat tracing can be guaranteed. In order to work properly, the CarnTrack concept requires strict adherence to this rule, thus placing a great deal of responsibility on the workers.

Box and carrier free traceability systems for meat products are already available on the market. One such system was developed by the Icelandic company Marel (2004). It assures traceability by

- guiding beef quarters to defined workstations, each consisting of a deboner and a trimmer,
- assigning predefined cutting tasks to these teams,
- requiring the trimmer to place exactly this piece of meat in a computer-controlled buffer and to confirm completion of this cutting task.

The tracking system then releases the meat from the buffer in a controlled way which makes tracking on the conveyor possible.

A completely different approach to assure traceability during the production process uses special printers to apply some kind of 'barcode' directly on the product (E+V, 2004). Although the ink used in these applications consists of food-compliant ingredients, the question is whether consumers will accept the negative impact of the method on product appearance. A practical problem to be solved is the often varying surface structure of food products, which may make it difficult to apply the code with the accuracy needed.

This section has shown that, in addition to traditional tracking techniques like barcode labels and RFID tags, several other methods exist to ensure product individual traceability. However, these methods either require relatively huge investments or have other drawbacks, so that their range of application is limited to a few specific scenarios. It can therefore be assumed that, even in the future, barcodes and RFID will remain the backbone of most traceability systems.

6.3.5   Data handling

So far, product identification and data acquisition have been discussed separately, without taking into consideration the need to transfer all information to a data processing system. Today, a wide range of communication technologies and standards is available to solve this task, including wireless alternatives (e.g. W-LAN, Bluetooth, GSM and UMTS) and robust bus systems (e.g. CAN-Bus, Profibus and JINI). As far as the software is concerned, many companies use a multi-layered approach, comprising production data acquisition systems (PDA), manufacturing execution systems (MES) and enterprise resource planning (ERP) systems. The various software layers in a production are shown in Fig. 6.4.

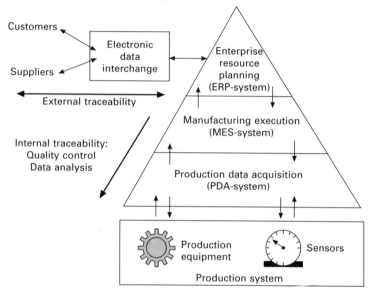

Fig. 6.4   Software layers in a manufacturing enterprise.

The production data acquisition system has the task of collecting and storing data from the machinery and the sensors on the production line. These data are then made available to the manufacturing execution system, which steers and coordinates the production process, taking the current state of the equipment into consideration. At the top level, ERP systems are 'accounting-oriented information systems for identifying and planning the enterprise-wide resources needed to take, make, distribute, and account for customer orders' (Hyperdictionary, 2004). To make the picture even more complex, companies have to exchange information with other members of the supply chain, for which the ERP system often uses the so-called electronic data interchange (EDI) standard. As can be seen, a large number of information systems are directly or indirectly attached to the manufacturing process. When implementing traceability concepts including process data, it would be helpful to use this existing infrastructure for system implementation. Traceability data can already be handled by many MES and ERP systems. However, standards for the exchange of food product data have not been developed yet. Nevertheless, examples from other industries show that such standardisation is both useful and feasible.

An example is the international standard for product model data (ISO 10303). This standard provides so-called 'integrated generic resources' (ISO 10303-41 to 49), that is 'information models which describe product data independent from a certain application' (ProSTEP, 2004), using the EXPRESS language (ISO 10303-11). In addition to this general framework, a great number of application protocols (AP) have been developed. An AP contains standardised descriptions of product data for a specific context. Each application protocol consists of the following components (ProSTEP, 2004):

- An application activity model (AAM), describing the functionalities of the IT solution for which the application model is specified. As a description method, the structured analysis and design technique (IDEFØ, see National Institute of Standards and Technology, 1993 and Pfleeger, 1998) is used, defining each application through processes with activities, inputs, outputs, controls and mechanisms.
- An application reference model (ARM), describing the IT solution from a user's perspective. For this purpose, a number of modelling techniques are available, the most widely used being EXPRESS and its graphical representation, EXPRESS-G.
- An application interpreted model (AIM), providing the mapping between the application reference model and the integrated generic resources defined in ISO 10303-41 to ISO 10303-49.

Existing protocols focus on applications such as 'Electrotechnical Design and Installation' (AP212) or 'Core Data for Automotive Mechanical Design and Processes' (AP214). None of these protocols, however, deals with food-related issues. Nevertheless, ISO 10303 provides a framework and modelling techniques which could serve as a basis for developing a data storage and exchange standard for traceability systems including process information. However, a lot of work still needs to be done, as several of the integrated generic resources focus on issues that are not relevant for the food processing industry (e.g. assembly processes).

6.4   Statistical methods for data analysis

If the traceability and data acquisition systems are running, a preliminary set of production data can be obtained by manufacturing an initial batch of products for test purposes. This batch is intentionally produced using varying processing parameters, in order to get an overview of product behaviour. After production, all products are examined and their microbiological, organoleptic and nutritional properties are evaluated. Products failing this test are classified as rejects, which will be stored in the database. For a more detailed analysis, the cause of the rejection should be stored, too. These initial data serve as a basis for the implementation of product evaluation models. At a later stage, during full-scale production, additional data sets will be added in order to improve the accuracy of the model.

There are several statistical methods which can be used for data analysis and classification. Two such methods will be presented in this section: SIMCA (Soft independent modelling of class analogy) and PLS-DA [partial least squares (or projection to latent structures) – discriminant analysis]. These methods can be used for classification of products into various groups, e.g. 'good taste/bad taste' or 'acceptable/unacceptable microbial count'.

6.4.1 SIMCA

The Soft Independent Modelling of Class Analogy works as follows (Joliffe, 2002):

- The data set is divided into exclusive populations.
- On the data set of each population, a principal component analysis (PCA) is performed (for more details on PCA, see Joliffe, 2002 and Næs et al., 1996). Applied to the process data, the principal component analysis constructs new variables such that
  - the new variables are linear combinations of the original ones, with the weight vectors that define the combinations being of unit length and orthogonal to each other
  - the new variables are uncorrelated
  - the first new variable captures as much as possible of the variability in all the original ones, having been constructed to have maximum variance amongst all such linear combination [and]
  - each successive new variable accounts for as much of the remaining variability as possible (Næs et al., 2002).

  The main benefit for the application discussed here is that the method delivers vectors of main variance in the n-dimensional parameter space. For each of these vectors, one can determine the percentage with which the vector contributes to the overall variance of the production and process data.
- In the next step, only those principal components (PC) will be retained which account for most of the variance in each population.
- The principal components retained can then be used to construct hyperplanes (or even less-dimensional subspaces), one for each population.
- The evaluation and classification of new data sets is then performed by measuring the distance from the new data point to these hyperplanes (or subspaces).

One possibility is to simply attribute the new product to the population whose subspace is closest, thus predicting product properties. However, it is also possible to define more sophisticated thresholds, e.g. for grouping the products into 'defective', 'non-defective' or 'unclear'. It is furthermore possible to define 'confidence volumes' around the two groups. A graphical representation of the method is shown in Fig. 6.5.

6.4.2 Partial least squares – discriminant analysis

A second method suitable for product classification using a multi-dimensional set of process data is the so-called partial least squares (or projection to

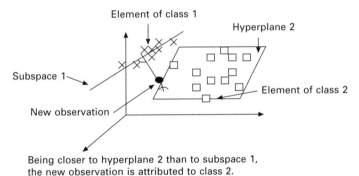

Being closer to hyperplane 2 than to subspace 1,
the new observation is attributed to class 2.

Fig. 6.5   An illustration of the SIMCA method.

latent structures) – discriminant analysis (PLS-DA). Based on a set of 'training data', this method creates a set of new input variables as linear combinations of the old input variables and then uses the new variables as predictors for output data (Wold *et al.*, 2004), that is, in this case, product quality. Contrary to SIMCA, which identifies local (PCA) models for each of the groups defined, PLS-DA 'is performed in order to sharpen the separation between the group of observations … such that a maximum separation between classes is obtained'. The goal is to understand which variables carry the class-separating information (Vinzi and Lauro, 2003). Like SIMCA, PLS-DA allows for the prediction of class affiliation of new observations. In the case of two classes, this is simply achieved by defining an affiliation variable ranging between zero and one (N N, 2004).

In general, both methods presented here work better if all classes under analysis are very tight, because this makes it easier to predict class affiliation. This is especially relevant for the PLS-DA, which fails if there is not enough homogeneity inside the classes (N N, 2004). On the other hand, PLS-DA provides useful additional information by identifying the variables responsible for class separation. While PLS-DA and SIMCA are able to solve tasks related to the classification of data, they do not establish correlations between (continuous) input and output parameters. For such purposes, other methods are available, such as partial least square regression (PLS).

6.5   Conclusions

As shown above, multivariate statistical methods can be used to implement flexible process parameter specifications by performing a holistic analysis of the production process. This is especially beneficial in cases where product properties cannot be monitored directly and where, as a consequence, process data are the only way for 100% product evaluation. In order to be able to perform the analyses and classifications discussed above, a link between the product and its processing data is essential, thus making the inclusion of process data in traceability a necessary prerequisite for the application of the method.

6.6   Future trends

Traceability issues have been very much in the focus of EU regulatory activities during recent years. However, the relevant regulation for the food industry, 178/2002/EC (EU, 2002), remains quite vague about how specific and detailed tracing information has to be. So far, it is only necessary to keep records of direct suppliers and customers. It is quite likely that these traceability requirements will be sharpened in the near future. The regulation has already established procedures for the adoption of provisions on how the overall traceability requirements shall be implemented [article 18(5)]. Thus, the issue of traceability will probably become even more important in the future than it is today.

As far as sensors are concerned, it is likely that even more devices will become 'real-time-capable', thus facilitating process data acquisition. This trend is, to a great extent, industry-driven, which means that many companies would like to test even more product factors continuously if suitable and affordable sensors were available.

Another trend which can be observed is that traceability solutions will become more affordable, thus making the tracing of process data available for an ever broader range of products. Current trends include the development of printable, organic integrated circuits (Clemens, 2004). In these circuits, conductive polymers are used to build electrical components such as transistors (Polymer field effect transistors, PFET). One application envisioned by the developers is to use these components for RFID tags. As the polymers can be printed, it will become possible to apply an RFID tag directly to a product's packaging without any specific integration needs. Production costs are predicted to be significantly lower than for silicon-based tags. However, the new technology is still under development and will take some years before being ready for mass production.

If and how far the integration of product-individual process data in traceability will play a role in everyday operations remains to be seen. This will greatly depend on the additional costs of such individualised traceability concepts, as operating margins for many food products are quite low and often leave little space for additional investments.

6.7   Sources of further information and advice

As this chapter covered an unusually wide range of topics, it was not possible to discuss every aspect in detail. Therefore, the interested reader is directed to websites and addresses where additional and more detailed information can be obtained for several of the issues discussed above. Please note that this list does not cover all suppliers and institutes working in the respective fields and that it should not be understood as a recommendation to buy products of the companies mentioned below.

On European regulations and directives:

Office for Official Publications of the European Communities, 2, rue Mercier, L-2985 Luxembourg, Tel: (352) 2929-1. This office publishes all EU regulations and directives and manages the website EUR-LEX (http://www.europa.eu.int/eur-lex/en/index.html) where all relevant documents can be downloaded free of charge.

On product and process data acquisition and analysis:

Matforsk AS, Norwegian Food Research Institute, Osloveien 1, N-1430 Ås, Norway. Tel: +47 64 97 01 00, Fax: +47 64 97 03 33. This institute is developing statistical tools for product and process optimisation (for the project homepage, see http://www.matforsk.no/web/wforsk.nsf/webTemaPE/200104!OpenDocument). Other research activities include the evaluation of food quality using rapid sensors for online analysis.

Manufacturers of software for multivariate analysis:

Infometrix Inc., 10634 E. Riverside Dr, Suite 250, Bothell, WA 98011, USA, info@infometrix.com. Infometrix develops software for the interpretation of patterns in multivariate data. An overview of applications of the company's products in the food industry, as well as additional information and references can be found on http://www.infometrix.com/apps/15-1096_FoodBevAO.pdf.

Eigenvector Research, Inc., 830 Wapato Lake Road, Manson, WA 98831, USA. Tel: ++1 509 687 2022, Fax: ++1 509 687 7033. Eigenvector research develops and distributes a toolbox for multivariate analysis to be used with Matlab™.

UMETRICS, Tvistevägen 48, Box 7960, SE-907 19 Umeå, Sweden. Tel: +46 90 184800, Fax: +46 90 184899. Umetrics is a manufacturer of software tools for data analysis, e.g. SIMCA tools.

On sensor selection and sensor integration:

Kress-Rogers E, Brimelow C J B (2001): *Instrumentation and Sensors for the food industry (2nd edition)*, Cambridge: Woodhead Publishing. A useful book containing very detailed information on numerous online sensor principles.

Fraunhofer IPA, Department 340, Nobelstr. 12, D-70569 Stuttgart, Germany. Tel: +49 711 970 1234, Fax: +49 711 970 1005, http://www.ipa.fhg.de/english/Arbeitsgebiete/BereichD/520/leistungsangebote/sensor/index.php. A research and development institute, dealing with sensor integration and applied sensor research (e.g. on fluorescence measurements).

Leatherhead Food International, Randalls Road, Leatherhead, Surrey, KT22 7RY, UK. Tel: +44 (0)1372 376761, Fax: +44 (0)1372 386228. A consultancy and research institution involved in numerous food research projects, including several sensor-related ones.

On data exchange standards and data specification:

ProSTEP iViP, Dolivostr. 11, D-64293 Darmstadt, Germany. Tel: +49-6151-9287-336, Fax: +49-6151-9287-326, E-mail: psev@prostep.com. The ProSTEP iViP association promotes development, utilisation and enhancement of STEP, the standard for the exchange of product model data. A lot of useful information about ISO 10303 can be found on the association's website: http://www.prostep.org/en/stepportal/was.

National Institute of Standards, Information Technology Laboratory, 100 Bureau Drive, Stop 8900, Gaithersburg, MD 20899-8900, USA. This institute published the IDEFØ standard. The whole standard can be downloaded from http://www.itl.nist.gov/fipspubs/idef02.doc.

International Organization for Standardization, Central Secretariat, 1, rue de Varembé, Case postale 56, CH-1211 Geneva 20, Switzerland. Telephone +41 22 749 01 11, Fax +41 22 733 34 30, www.iso.org. Publishes and distributes the standards from the ISO 10303 series.

6.8   Bibliography

Benson, I B and Millard, J W F (2001), 'Food compositional analysis using near infrared absorption technology', in Kress-Rogers E and Brimelow C J B (Eds.), *Instrumentations and Sensors for the Food Industry*, Cambridge, Woodhead Publishing, 138–186.

Clemens, W (2004), Preiswerte Funketiketten mit gedruckter Polymerelektronik, in Bey I, Karlsruher Arbeitsgespräche 2004 – Wege zur Individualisierten Produktion, Karlsruhe, FZK.

Dowell, F E, Ram, M S and Seitz, L M (1999), *Cereal Chem.*, 76(4), 573–576.

E+V Technology GmbH (2004), VCIS 2000: System zur lückenlosen Identifizierung und Nachverfolgung von Schlachttierkörpern, http://www.eplusv.de/start.htm.

EU (1985), Council Directive 85/374/EEC of 25 July 1985 on the approximation of the laws, regulations and administrative provisions of the Member States concerning liability for defective products, Brussels, Office for Official Publications of the European Communities.

EU (1999): Directive 1999/34/EC of the European Parliament and of the Council of 10 May 1999 amending Council Directive 85/374/EEC on the approximation of the laws, regulations and administrative provisions of the Member States concerning liability for defective products, Brussels, Office for Official Publications of the European Communities.

EU (2002), Regulation (EC) No 178/2002 of the European Parliament and of the Council of 28 January 2002 laying down the general principles and requirements of food law, establishing the European Food Safety Authority and laying down procedures in matters of food safety, Brussels, Office for Official Publications of the European Communities.

Farkas J (2003), *Rapid detection of microbial contamination*, Paris, Fair-Flow/INRA.

Food Manufacturing Coalition for Innovation and Technology Transfer (1997): 'Sensors for Inspection, Sorting, Grading, Process Monitoring and Control', Great Falls http://foodsci.unl.edu/fmc/9sensor.pdf.

Ford Motor Company (1996), Failure Mode and Effect Analysis – Handbook Supplement for Machinery, FMC, http://www.fmeainfocentre.com/download/fordmachineryfmea.pdf.

Harris, C (2004), 'Dumeco – integrated logistics along the line', *Meat Processing Global*, 9/10, 12–17.

Huen, E, Huen, J and Stallkamp, J (2003), Verfahren zur kontrollierten Erfassung des Bearbeitungsweges wenigstens eines Teils eines Stückguts sowie ein diesbezügliches Erfassungssystem, Offenlegungsschrift DE 101 32 647 A1, München, DPMA.

Huen, E and Klafft, M (2004), Gewährleistung der hygienischen und organoleptischen Qualität der Lebensmittelprodukte, in N.N., Food Traceability, Stuttgart, Fraunhofer IPA (CD-ROM).

Humber, J (1993), Kontrollpunkte und kritische Kontrollpunkte, in Pierson, Corlett (Eds.): HACCP – Grundlagen der produkt- und prozeßspezifischen Risikoanalyse, Hamburg, Behr's Verlag, 123–131.

Hyperdictionary (2004): http://www.hyperdictionary.com.

ISO 10303-1:1994 Industrial automation systems and integration – Product data representation and exchange – Part 1: Overview and fundamental principles.

ISO 10303-11:1994 Industrial automation systems and integration – Product data representation and exchange – Part 11: Description methods: The EXPRESS language reference manual.

ISO 10303-41:1994 Industrial automation systems and integration – Product data representation and exchange – Part 41: Integrated generic resources: Fundamentals of product description and support.

ISO 10303-42:1994 Industrial automation systems and integration – Product data representation and exchange – Part 42: Integrated generic resources: Geometric and topological representation.

ISO 10303-43:2000 Industrial automation systems and integration – Product data representation and exchange – Part 43: Integrated generic resources: Representation structures.

ISO 10303-44:2000 Industrial automation systems and integration – Product data representation and exchange – Part 44: Integrated generic resources: Product structure configuration.

ISO 10303-45:1998 Industrial automation systems and integration – Product data representation and exchange – Part 45: Integrated generic resource: Materials.

ISO 10303-47:1994 Industrial automation systems and integration – Product data representation and exchange – Part 47: Integrated generic resource: Shape variation tolerances.

ISO 10303-49:1998 Industrial automation systems and integration – Product data representation and exchange – Part 49: Integrated generic resources: Process structure and properties.

ISO 10303-212:2001 Industrial automation systems and integration – Product data representation and exchange – Part 212: Application protocol: Electrotechnical design and installation.

ISO 10303-214:2003 Industrial automation systems and integration – Product data representation and exchange – Part 214: Application protocol: Core data for automotive mechanical design processes.

Joliffe, I T (2002), *Principal component analysis (2nd edition)*, New York, Springer.

Kent, M (2001), Microwave measurements of product variables, in Kress-Rogers and Brimelow (Eds.): *Instrumentations and Sensors for the food industry*, Cambridge, Woodhead Publishing, 233–279.

Kilcast, D (2001), Modern methods of texture measurement, in Kress-Rogers and Brimelow (Eds.): *Instrumentations and Sensors for the food industry*, Cambridge, Woodhead Publishing, 518–549.

Kuhn, C and Huen, J (2004), *Lebensmitteltechnik*, **7-8**, 66–67.

Leistner, L and Gould, G W (2002), *Hurdle Technologies: Combination Treatment for Food Stability*, New York, Kluwer Academic/Plenum Publishers.

Marel (2004), From breeder to buyer – traceability all the way, http://www.marel.com/11000/11400_case.asp.

National Institute of Standards and Technology (1993), Draft Federal Information. Processing Standards Publication 183, 1993 December 21, Announcing the Standard for INTEGRATION DEFINITION FOR FUNCTION MODELING (IDEF0), Gaithersburg.

Næs, T, Baardseth, P, Helgesen, H and Isacsson, T (1996), 'Multivariate Techniques in the Analysis of Meat Quality', *Meat Science* **43(S)**, 135–149.

Næs, T, Isaksson, T, Fearn, T and Davies, T (2002), *A User-friendly Guide to Multivariate Calibration and Classification*, Chichester, NIR Publications.

N N (2004), PLS_Toolbox 3.5 for use with MathLab™, Manson, Eigenvector Research.

Pfleeger, S L (1998), *Software Engineering – Theory and Practice*, Upper Saddle River, Prentice Hall.

ProSTEP iViP Association (2004), Application Protocols, http://www.prostep.org/en/stepportal/was/ap.

Ridley, I (2001), Practical aspects of infra-red remote thermometry, in Kress-Rogers E and Brimelow C J B (Eds.), *Instrumentations and Sensors for the food industry*, Cambridge, Woodhead Publishing, 187–212.

Roberts, I (2001), In-line and on-line rheology measurement, in Kress-Rogers E and Brimelow C J B (Eds.), *Instrumentations and Sensors for the food industry*, Cambridge, Woodhead Publishing 2001, 403–422.

Schmilovitch, Z, Hoffman, A, Egozi, H, Nelson, S O, Kandala, C V K and Lawrence, K C, 1996 *Appl. Eng. Agric.*, **12(4)**, 475–479.

Tsenkova, R, Atanassova, S, Toyoda, K, Ozaki, Y, Itoh, K and Fearn, T (1999), 'Near-Infrared Spectroscopy for Dairy Management: Measurement of unhomogenized Milk Composition', *J. Dairy Sci.*, 1999, **82**, 2344–2351.

Vinzi, V E and Lauro, C (2003), PLS Regression and Classification, Lisbon, PLS'03.

Waldner H (2004), Die EU-Lebensmittel-Basisverordnung, in: Fraunhofer IPA: Food Traceability – Qualitätskontrolle unter der EU-Verordnung 178/2002, Stuttgart: FH-IPA.

Wold, S, Eriksson, L, Trygg, J and Kettaneh, N (2004), The PLS method – partial least squares projections to latent structures – and its applications in industrial RDP (research, development and production), Prague, COMPSTAT.

7

Traceability of analytical measurements

M. F. Camões, University of Lisbon, Portugal, and R. Bettencourt da Silva, Directorate General for Crop Protection, Portugal

7.1 Introduction: the role of analytical measurements in evaluating product quality

The assessment of the composition of food products of various kinds and origins has been considered greatly important by various civilizations throughout history. Since the XIVth century, concerns about public health and protection of the economy have led to legislation in European countries relating to the quality of foodstuffs, additives and drinks. Nevertheless, understanding the physical and chemical phenomena involved in those evaluations was a possibility only after W. Ostwald (1853–1932) introduced the Arrhenius (1859–1927) theory of electrolyte solutions to qualitative analysis, thus defining the early stages of modern analytical chemistry, one century after Lavoisier (1743–1794) promoted chemistry to an exact science through the introduction of the law of conservation of matter.

Nowadays, the importance of the role of chemical measurements in food safety and in regulation of the food trade is unquestionable. In an industrial society, the cost of chemical characterization of food products can reach 3% of the gross income or, when there is need to repeat the analysis for confirmation of a result, it can easily become three times larger[1]. The control of food safety allows evaluation and management of consumer exposure to physical, chemical or biological entities that could affect health, and stimulates the production and trade of safe foods. Since 'National Official Monitoring of Food Products' can only control a small fraction of the consumption, the potential severe economic impact of publicity of unsafe foodstuffs motivates trade partners to guarantee the safety of these products. Furthermore, the evaluation of the capability of the food product to satisfy consumer needs and expectations, beyond its safety, is extremely important to protect the economic interests involved in the commercialization of these goods. The concern of the foodstuffs producers and traders in protecting their economic interests has meant that most of the foodstuffs quality control is performed or contracted out by the trade partners, in order to protect their credibility in the market or to define the product's commercial value. The globalization of trade and the associated diversification of trade relations and, therefore, the need for transparency and safety of the food transaction are forcing trade partners to control even more carefully the characteristics of the transacted goods.

The transparency of foodstuffs quality control can be guaranteed by the analysis of the products by laboratories, whose independence and competence is ideally evaluated by a third party having recognized competence for laboratory evaluation. Accreditation[2] of test laboratories by Accreditation Bodies whose competence is directly or indirectly recognized by the 'International Laboratory Accreditation Cooperation' (ILAC), based on the ISO/IEC 17025 Standard[2], allows the recognition of the competence of laboratories from all over the world[3]. However, many food analyses are performed in laboratories whose competence in not easily proved.

The analytical study of products in order to gain knowledge of their composition, origin and purity also impinges on the technical optimization of agricultural practices, production and processing, handling, storage and preservation. The possible contamination of food by environmental pollutants that may be harmful to health, establishes close links with environmental problems; thus, the analysis of food and foodstuffs covers a very broad field of scientific and technical issues.

The technical quality of food characterization depends on the adequacy of the performed measurements to objectively answer the question posed that constitutes the analytical problem. The degree of excellence in the (bio)chemical information supplied with a view to solving the analytical problem encompasses four components: quality of results, quality of the process, quality of the analytical tools used and the quality of the work and its organization. In order to ensure Quality, laboratories performing tests need to implement Quality Control (QC)[4]. The recognition of performing laboratories as independent and competent entities implies external Quality Assessment, ideally evaluated and ensured, Quality Assurance (QA), by a third party of recognized competence.

Considering the characterization of foodstuffs, the analytical problem can be for instance: 'Is the food product contaminated with harmful or legal levels of lead?' or 'Has the food product the origin or the value claimed by the trader?' The first example involves comparison of the foodstuffs lead content with a certain pre-defined level whereas the second example involves comparison of the sample composition with a pattern that characterizes the composition of products from a stated origin. The answer to the analytical problem can only be adequate if the generated analytical result is comparable with the reference of the evaluation, i.e., if the measured quantities are exactly the ones referred to in the legislation or in the product specification.

Knowledge of the quality of the analytical result that allows estimates of the possible differences between the analytical result and the food 'true' characteristic is essential for its comparison with a reference. An example of the inadequacy of the traceability of a result for the characterization of a food product is given by the difference between the definition of the legal maximum content of mercury allowed in fish, and the results from the analysis of this metal in the food product considering a method of analysis that recovers only 80% of the mercury content in the fish due to uncorrected losses of the metal by volatilization. In this particular case, the exposure of the consumer is underestimated by the fact that the laboratory is measuring only the mercury that can be recovered by the analytical method and not the 'real' content of the metal in the fish that will contaminate the consumer and that is referred to in the legislation. One way to clarify this difference between the measurement result and the reference value is to claim that neither value is traceable to a common reference.

Comparability of analytical results with each other and with a legislated or specified value is only possible if the results are traceable to the reference of the evaluation or to a reference of higher level to which the reference of evaluation is referred, i.e, traced. The traceability of the result of a measurement or of the value of a standard is defined as metrological traceability, the property of the result of a measurement or the value of a standard whereby it can be related to stated references, usually national or international standards, through an unbroken chain of comparisons all having stated uncertainties. This is distinct from documental traceability or product traceability that refers, respectively, to the knowledge of the origin and the path followed by a document or a product, extensively discussed in the other book chapters. Traceability is essential for comparability of analytical results and is a requirement of the ISO/IEC 17025 standard[2].

A requirement for the comparability of the information reported by the laboratory with the evaluation reference or with another result, is the expression of results with an objective measure of their quality. This is usually fulfilled through the expression of results with a quantity that estimates the difference between the best estimation of the result, i.e., the single value that is reported by most of the test laboratories, and the 'true value' of the measured quantity considering the reference of the measurement to which the measurement result is traced. This quantity is internationally defined as uncertainty, a parameter associated with the result of a measurement, that characterizes the dispersion of the values that could reasonably be attributed to the measurand or quantity subjected to measurement, and should be reported with a certain confidence level, usually 95 or 99%, that states the probability of that uncertainty estimating the difference between the best estimation of the result, and the 'true value', which is internationally defined as 'measurement error'. Some authors prefer to use the term 'internationally accepted value' instead of 'true value' due to the fact that the truth is usually unknown or unattainable. The analytical result is then presented in the form of an interval around the central value which constitutes the best estimation of the result.

The definitions of traceability, measurement uncertainty and measurand[4] reflect the close relationship of these three concepts. The adequate definition of the measurand is important to define the traceability of the measurement result necessary to solve the analytical problem, and the standards needed for the analysis. The definition of metrological traceability considers a chain of comparisons of measurement results or standards because in some cases both the estimated measurement results and the standards used are not referred directly to the top of the traceability chain, i.e., they are compared with items directly or indirectly referred to the traceability chain. All the performed comparisons must be performed with known uncertainty allowing the estimation of the uncertainty of the measurement result or standard value.

7.2   Problems in tracing and comparing analytical measurements

7.2.1   Traceability of measurements results to the SI units

Ideally, the traceability of all results from the characterization of foodstuffs should be established in a way equivalent to the one already developed for many physical measurements[5]. Nowadays the comparability, all over the world, of the results from measurements of several physical quantities, such as length and mass, is based on a traceability chain constituted by standards of different metrological quality managed by a network of international physical laboratories which ensure that each of the used measurement references are related to a common reference, the primary standard, with a known uncertainty. This procedure allows the measurements performed under this umbrella to be also traceable to an international reference.

The need for comparability of the more frequent physical measurements performed throughout the world was identified in the middle of the 19th century. The first universal exhibition, a great exhibition of the Works of Industry of All Nations which opened on the first of May 1851 in the Crystal Palace in London organized by the Society for the Encouragement of Arts, Manufactures and Commerce and realized by Sir Henry Cole, curiously enough famous for his phrase 'Learn how to see and see through comparison', can be traced as the landmark for developments that followed. In 1875, a diplomatic conference took place in Paris where 17 governments signed a treaty, 'The Metre Convention', by which it was decided to create and finance a scientific and permanent organizational structure for member governments to act in common accord on all matters relating to units of measurement. The 'Bureau International des Poids et Mesures' (BIPM)[6] was founded in order to design a system of units to be used throughout the world; it was given authority to act in matters of world metrology, particularly concerning the demand for measurement standards of ever increasing accuracy, range and diversity and the need to demonstrate equivalence between national measurement standards. At the beginning of 2005, 51 states were members of the Metre Convention. In 1946, the Metre Convention countries accepted the MKSA system (metre/m, kilogram/kg, second/s, ampere/A), that was extended, in 1954 to include the kelvin/K and candela/cd. This system was named 'The International System of Units', SI (Le Système International d'Unités) in 1960. Nowadays, the International System of Units is comprised of seven dimensionally independent units, called Base Units, the previous ones and the mole/mol, and 16 units called Derived Units resulting from the multiplication and division of Base Units, (e.g. the derived unit for density, kilogram per cubic metre, kg/m^3). The BIPM is organized in thematic technical consulting committees, one of them, created in 1993, being the Comité Consultatif pour la Quantité de Matière, CCQM, whose present activities concern primary methods for measuring amount of substance, and international comparisons, establishment of international equivalence between national laboratories, and advice to the BIPM on matters concerned with metrology in chemistry[6].

Although there is a Base Unit for the amount of substance, the mole or mol, that represents the usual target of chemical analysis, normally the results from chemical measurements are not traceable to the SI units. This is because the traceability chain, that supports the measurement results comparability, needs more than conceptual evolutions. The material existence of adequate standards or measuring methods is vital to perform chemical measurements of the highest metrological quality. The tremendous diversity of chemical analyses, where, for instance, the methodologies used for the analysis of lead in water, in a vegetable, or in an alloy, are totally different, and the complexity of most of the chemical measurement processes, that can present some serious weaknesses regarding the presence of various types of interferences, makes it extremely difficult to construct a hierarchical system of measurements and standards equivalent to the one established for some physical measurements, Fig. 7.1.

However, there are some chemical measurements that can claim traceability of their results to the SI units, shorthand for 'traceable to reference values obtained by agreed realizations of the SI units'. Figure 7.2 presents an example of the structure of a hierarchical system of standards and methods that can support the traceability of chemical measurements to the SI units. The construction of such a traceability chain needs the existence of pure chemical standards and primary methods of analysis that are positioned at the top of the traceability chain. A primary method of measurement is a method having the highest metrological qualities, whose operation can be completely described and understood, for which a complete uncertainty statement can be written down in terms of SI units. A primary direct method measures the value of an unknown without reference to a standard of the same quantity. A primary ratio method measures the value of a ratio of an unknown to a standard of the

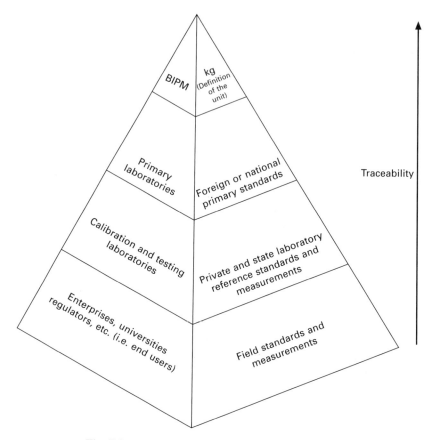

Fig. 7.1   Traceability chain for weight measurements.

same quantity; its operation must be completely described by a measurement equation[7].

In some fields of analytical chemistry, e.g. in the characterization of organic contaminants, such as pesticides in foodstuffs, there is lack of chemical standards having the sound purity statements necessary to guarantee the traceability of measurement results to the SI units. Primary methods are methods whose principles and quality of performance guarantee the production of results of the highest quality upon pure chemical standards. Those methods are further used to validate and calibrate secondary methods, directly through method comparison performed analysing the same samples by both methods, or indirectly through the certification of primary reference materials that can be analysed by the second level of methods of analysis. Secondary methods are then directly or indirectly used in quality assurance programmes developed to evaluate the performance of the methods of analysis currently used in test laboratories. Usually, the primary, secondary and routine methods of analysis are successively, in this order, cheaper, simpler, faster and more uncertain. The primary methods of analysis must be capable of performing measurements free from interferences of any kind and with a low measurement uncertainty. The CCQM identified the following measurement methods as being capable of supporting primary methods of analysis:

- isotope dilution with mass spectrometry (IDMS);
- coulometry;
- gravimetry;
- titrimetry
- determination of freezing-point depression; and
- differential scanning coulometry.

These measurement methods can only produce primary methods of analysis if the remaining analytical steps have the quality fit for that purpose, i.e., if they allow the production of a method with the characteristics of a primary method. Furthermore, the primary methods of analysis can only generate high-quality information in laboratories organized in order to achieve the excellence of the analytical work. One of the cornerstones of the production of results of the highest quality is the development of analytical method validation and analysis quality control schemes dimensioned for this goal.

All stages of the traceability chain, presented in Fig. 7.2, must be managed by different levels of laboratories and must be performed with a known

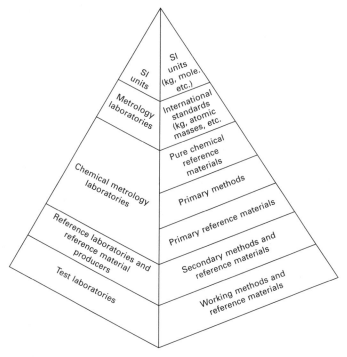

Fig. 7.2   Example of the hierarchy of methods and standards capable of ensuring the traceability of a chemical measurement result to the SI units.

uncertainty that quantifies the strength of the chain links. The highest metrological capabilities, i.e. the Primary methods and/or substances, are the scope of action of the National Metrology Institutes, NMIs, which have a more or less broad range of capabilities. It is the competence of the BIPM/ CCQMs to assess those Calibration and Measurement Capabilities (CMCs), namely those on Amount of Substance/Quantité de Matière, QM; the chemical categories covered are: high-purity chemicals; inorganic solutions; organic solutions; gases; water; metal and metal alloys; advanced materials; biological fluids and materials; food; fuels; sediments, soils, ores and particulates; and other materials.

The greatest advantage of achieving the traceability of results to SI units is to guarantee that the generated data reaches a level of quality that allows its comparability with data generated, with the same care, all over the world and through time, i.e., in the past or in the future, independently of the methodologies used. This feature is extremely important, for instance, when studying the evolution of the contamination of food with chemicals accumulated in the environment. Considering the important difficulties in producing chemical measurements traceable to the SI units, analytical chemists developed alternatives to the construction of a complex hierarchical system of methods and standards to support the comparability of the measurements; a vital feature of any kind of measurement. Taking into account the aim of the analysis, when necessary and possible, the comparability of measurements is preserved even when there is lack of material objectivity of the measurand, that is of the measured quantity.

7.2.2   Traceability of measurement result to the used method and standard

In chemical measurements and, in particular, in food analysis, the generated results are often not traceable to the SI units. In those situations, the comparability of measurement results is based on a shorter traceability chain. This can be exemplified with the case of the determination of protein content in foodstuffs, based on the quantification of total nitrogen and on the conversion of that value to protein content using conversion tables for several kinds of food products. The results from these analyses are a function of the principles of the method which simplifies the difficult task of estimating the mass fraction of all compounds chemically defined as proteins. Therefore, the result from the protein content estimated by one particular laboratory, considering this methodology, is only comparable to the result produced by another laboratory that uses the same analytical method. In this case, the measurand is the protein content defined by the method principles and assumptions, and the obtained results are traceable to the analytical method that constitutes the reference of the measurement. Since the analytical method under concern does not quantify an objective chemical entity and proposes an experimental procedure to estimate a parameter that is difficult to describe chemically and materially, it is defined as an empirical method. Other examples of empirical methods include those for acid content, fat content, peroxides index.

Many of the more important characteristics of foodstuffs are evaluated using empirical methods of analysis because of the difficulty in describing chemically such complex products. In situations in which the quantified entity is very well defined in chemical terms, the methods of analysis employed are classified as rational. Examples of these latter methods are the determination of amino acids and of lead in foodstuffs, for which the analysis is usually set on the use of standards of the same chemical species as that under quantification in the foodstuffs. In such cases, the laboratory can use any method of analysis capable of estimating the analyte (the chemical species whose concentration in the sample is to be determined) content in the sample. When the method used does not estimate the analyte adequately, the results are not comparable with the ones generated by a correct analytical methodology and the laboratory is producing results traceable to the method instead of traceable to a value of higher order. When there are no adequate reference materials, characterized by primary or secondary methods, to assess the performance of the method used, it is difficult to known the quality of the performed measurements and it is more correct to claim the traceability to the method alone.

In some fields of analysis, e.g. in the estimation of the content of organic contaminants in foods, the low quality of the standards used in the calibration of the methods of analysis can also be a serious obstacle for the comparability of results generated using standards from different manufacturers. In some analytical fields, the standards purity is estimated using techniques that have known weaknesses. For instance, the purity of 'pure' pesticides is, in many cases, estimated by gas chromatography using a flame ionization detector and considering the ratio between the area of the peak corresponding to the pesticide under concern and the sum of the areas from all peaks of the chromatogram.

This methodology, usually called the '100% Method', is based on the principle that the signal obtained from this detector is proportional to the mass of the organic compounds giving a correct estimation of the mass fraction of the pesticide. However, in some cases, this assumption is not correct. When the organic compounds are likely to degrade at the high temperatures in a gas chromatograph, this instrumentation may be substituted by a more selective one, e.g. liquid chromatography with UV/Vis spectrometric detection, that uses the principles of the 100% Method with a strong probability of producing incorrect standard purity values.

Therefore, in the situation where the principles of the methodology used to estimate analyte content of the standards are weak, the standards are only comparable if the same methodology for the assessment of their purity and the same procedure for synthesizing and purifying the analyte in the standard are used; this means that it is only possible to guarantee the comparability of standards produced by the same company.

When pure chemical standards are used with purity defined by the manufacturing practice of the producer, the measurement results estimated by an analytical method calibrated with these references are also traceable to those references. When the laboratory uses an empirical method and at a certain stage uses a standard with a purity relevant to the test result, defined by the producer, the measurement result is traceable to the method and to the pure standard; that means that its results are only comparable with the ones produced by a laboratory that uses the same (or a comparable) method and standard.

In some cases of the analysis of organic matrices, such as in food, the analytical method performance considering reference samples artificially prepared in the laboratory, by adding analyte to the matrix (usually designated as 'spiking' to produce 'spiked' matrixes), can vary compared with the samples usually characterized by the laboratory (usually designated as 'incurred' samples), becoming a serious barrier to results comparability. The analyte can be strongly bound to the matrix in incurred samples reducing the analyte recovery observed for spiked samples. In this case, if two laboratories use significantly different procedures to prepare the reference samples, considering the difference between method performances for incurred and spiked samples, the two laboratories can have different perceptions of the method behaviour for real samples (i.e., incurred samples). This difference can be expressed by considering differently the pertinence and magnitude of the correction of the method accuracy. This problem can occur even if the two laboratories use the same method of analysis.

In analytical areas where there is a suspicion of the variation of the method performance with the analyte origin and adequate reference materials are not available, the comparability of the results from laboratories in the same sector can be evaluated quantitatively in a proficiency test (PT). Proficiency testing schemes (PTs) are regularly organised interlaboratory comparisons to determine the technical competence of laboratories, which test the same or similar items. PTs are excellent tools for determining the technical competence of laboratories and the comparability of analytical results produced by different laboratories[8].

In cases where the laboratory performance is only considered to be adequately evaluated in proficiency tests and when the laboratory produces satisfactory results, the results subsequently generated are traceable to the PT. This situation is independent of the reference value of the analyte characterized in the item through the proficiency test, being estimated either by the combination of the results from all the participants in the test, or by a competent reference laboratory. In the first case, the quality of the method performance evaluation should increase with the number of methods and method types (e.g. the determination of water hardness can be estimated volumetrically or by atomic absorption spectrometry, two completely different methodologies certainly prone to be affected by different interferences). In the second case, the quality of the method performance evaluation improves with the quality and traceability of the measurements performed by the reference laboratory. The analysis of pesticide residues in vegetables is an example where the comparability of the results is supported by the traceability of results to proficiency tests since different analytical methodologies, expectedly with different performance, are used.

Laboratories often have the opportunity of participating in proficiency tests organized at national or international level. A well known European database of existing PTs is EPTIS[8]. Examples of laboratory intercomparison exercises in the area of food are among those of the International Measurement Exercise Programme, IMEP[9], namely those about lead content in wine[10] and trace elements in rice[11].

When a laboratory has available Reference Materials or Certified Reference Materials equivalent to the incurred samples and uses them in the evaluation of the adequacy of the method to perform measurements, the generated results are traceable to that reference material. Limited availability of appropriate Reference Materials is a constraint to many analytical determinations, especially for the analysis of chemical species in samples of very complex matrix, food usually being in that category. Development of new and better reference materials is a must and this is a task which is being undertaken by renowned organizations. This is the case, for instance, of the Institute for Reference Materials and Measurements (IRMM) of the European Commission[9] which produces and commercializes the European Reference Materials. When a sample is assessed with alternative tools, the one that should be claimed firstly to be the reference to the performed analysis is the one positioned at the higher level of the traceability chain.

Food test laboratories usually characterize a sample from a food lot that was built according to legislated rules or specifications. In these circumstances, the laboratory should state that the analysis results cannot be extrapolated to the sampled population, as considered in the ISO/IEC 17025 standard[2]. However, there are cases where those interested in the analytical result want to see the mean content of the analyte quantified in a lot of several tons of foodstuffs. In these cases, the heterogeneity of the lot must be estimated since it will affect significantly the measurement uncertainty value; it is often the source of the highest contribution to the final result uncertainty.

7.2.3   Estimation of the measurement uncertainty

Independently of the target for the measurement, knowledge of the measurement uncertainty is essential in order to support the traceability of the result. Without this objective measure of the result quality, it is not possible to claim the traceability of the generated analytical information to a certain reference.

In the analysis of foodstuffs, there are several examples of difficulties in the generalization of the estimation of the measurement uncertainty. These difficulties are usually not due to the incapacity of the developed methodologies to adequately estimate the measurement uncertainty, but rather to the lack of a comprehensive bibliography which hinders the understanding of the state of the art in this area.

In the last decade, several approaches were developed for the quantification of the measurement uncertainty. The main difference between these approaches resides mainly in the type of information on the analytical method performance that is used to estimate an important fraction of the measurement uncertainty. Independently of the used methodology all of them involve the same five steps[12,13]:

1 Definition of the measurand.
2 Identification of the sources of uncertainty.
3 Quantification of the sources of uncertainty.
4 Combination of the sources of uncertainty.
5 Calculation of the expanded uncertainty (i.e., of a high confidence level uncertainty).

Although the quantification and combination of the sources of uncertainty can involve complex calculations, the definition of the measurand and the qualitative identification of the sources of uncertainty are the more complex and important stages for a realistic estimation of the measurement uncertainty. The identification of the sources of uncertainty is commonly supported by the construction of Ishikawa diagrams, also called cause–effect diagrams, which avoids double counting of the sources of uncertainty. These diagrams can become very complex as the method complexity increases. Figure 7.3 shows a cause and effect diagram[21] for the determination of pesticide residues in vegetables.

Currently, four types of methodologies to estimate the measurement uncertainty can be identified:

- The ISO guide for the expression of the results with uncertainty[12] presents one usually designated as a bottom-up approach, based on the quantification of the quality of measurement through the combination of the uncertainty associated with each of the various fractions of the method that affect the quality of the produced information. This includes the uncertainties associated with both the performance of the analytical steps and the quality of the used chemical references. The uncertainties associated with these sources are, usually, combined through the uncertainty propagation laws. This approach is known as bottom-up, since it uses information from the performance of fractions of the method to estimate the uncertainty at a higher level, i.e. the information flows up. This approach should only be applied to analytical methods which involve sources of uncertainty that can be objectively and separately estimated before their combination. However, in food analysis, there are often analytical steps whose performance cannot be easily estimated separately, e.g. solvent extractions or acid digestions.
- In 1995, the Analytical Methods Committee of the Royal Society of Chemistry, presented a methodology[14] for the quantification of the

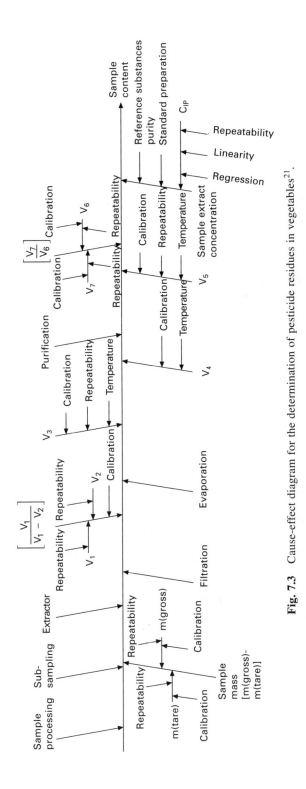

Fig. 7.3 Cause-effect diagram for the determination of pesticide residues in vegetables[21].

measurement uncertainty that does not need the subdivision of the analytical method in its many sources of uncertainty. This approach is based on interlaboratory information that transforms all the types of errors involved in the analysis, performed by each one of the individual laboratories, into random errors that can be easily quantified by experimental standard deviations. In some cases, the dispersion of results from the interlaboratory assay must be combined with other sources of uncertainty in order to extend the measurement uncertainty to other analytical steps such as processing of solid samples, or to different concentration levels. This approach was originally called top-down because it uses information from a higher level to estimate the quality of the information generated in intralaboratory environments; in this case, the information flows down. Nowadays this approach is more exactly designated as 'top-down based on interlaboratory information'. The most important drawback of this approach is its high cost and, in some cases, the limited scope of interlaboratory information.

- Considering the serious limitations of the previous approaches, several approaches have been developed[14–18] that use the information usually gathered during in-house validation, i.e., intralaboratory validation of the methods performance to estimate the measurement uncertainty. These approaches are also classified as top-down because they use information collected in a larger scope (e.g. extended period of time) than the one necessary to characterize a sample. Usually the intralaboratory validation of the method is performed in intermediate precision conditions (i.e., in an extended time interval, by several operators and using different material and equipment) and the assays are usually performed in repeatability conditions (i.e., by one operator, over a short period of time and using the material and equipment necessary for the assay). There are several examples of these approaches that use Analysis of Variance, ANOVA[17] or Factorial Analysis[18] in order to better understand the sources of uncertainty that can be optimized in order to improve the quality of the measurements or reduce the cost of analysis. However, the capabilities of these approaches to break up the method into its components and understand them, are far from those of the 'bottom-up' approach.

- More recently,[19–21] a new approach was developed for the quantification of the measurement uncertainty that gathers the more relevant advantages of the approaches described previously. This approach is designated as a 'differential' approach and involves first the division of the analytical method into analytical steps whose uncertainty can be estimated by well known models, called simple analytical steps (e.g. gravimetric and volumetric steps), and analytical steps, designated as complex, whose contribution to the method uncertainty is previously unknown due to the lack of models capable of describing their performance. Secondly, the performance of the complex analytical steps is estimated, by difference, through the comparison of the combination of the uncertainty associated to the simple analytical steps with global method performance parameters estimated from the intralaboratory validation of the method. Furthermore the estimated sources of uncertainty are combined, through the uncertainty propagation laws, in a detailed model of the method performance that can be validated for a certain concentration range. Since this approach, which allows the collection of additional information regarding the performance of the analytical steps, is based on information positioned above and below the level where the analytical result is produced, it can be classified as a 'transversal' approach.

7.3 Improving comparability of analytical measurements

The comparability of results from the analysis of foodstuffs, as in other fields, is only possible when the measurement results are traceable to an adequate reference and if they are presented with uncertainty, i.e., with an objective measure of their quality. Considering the definition of the traceability of the measurements, the comparability of the measurement results is supported by different cornerstones depending on the type of analytical methods used.

When the methods used are empirical and, therefore, all the laboratories use the same analytical method, the comparability of results can be affected by the incomplete and inadequate description of the method. In some cases, the quality of the chemical references used to calibrate the analytical instrumentation can also affect the analytical equivalence of the data generated by different laboratories. Interlaboratory assays, called collaborative assays, that involve the use of the same analytical method protocol, in order to evaluate the adequacy of the formal description of the analytical procedure, are powerful tools to find out if the empirical method can support the production of comparable results. One of the missions of the analytical method protocol is the description of the quality of the chemical references used. When the results from the collaborative assay are very dispersed, the adequacy of the principles and the description of the analytical procedure should be evaluated.

Each time rational analytical methods are used, the comparability of the results produced by different laboratories, using different methodologies, must be supported by the quality of the chemical references used. The quality of these tools becomes more and more important as the diversity of the principles of the currently-used analytical methodologies as well as the complexity of the studied analytical system increase. When the used rational methods behave differently considering the used chemical references (e.g. spiked samples) and the characterized items (i.e., incurred samples) the comparability of the results can be much affected. The promotion of proficiency tests in those analytical areas allows monitoring of the comparability and adequacy of the generated data to fulfil clients' needs. When there is evidence that the comparability of analytical measurements is not satisfactory in a field managed by rational methods, this drawback can be overcome by the production of adequate chemical references. When the production of these chemical references is not possible, for instance due to the large number of analyte/matrix combinations, the comparability of measurements can be supported by the generalized use of a reference method. In this case, the laboratories are obliged to base the measurement on the use of a reference method or equivalent. In this case, the measurements results are traceable to the reference method.

The option to trace the results to proficiency tests is not always the best choice since it only guarantees the comparability of results produced by the participating laboratories, for the selected and used reference samples. In these cases, if the analyte from the proficiency test sample is more available to the analysis than it is in incurred samples, e.g. if is not bound to the sample matrix, the results from these interlaboratory assays, can give a wrong perception of laboratories results comparability concerning the samples currently characterized in the laboratories; analysis of organic contaminants is full of such examples, such as pesticides in food.

Considering the quantification of the measurement uncertainty, the comparability of the estimation of the measurement quality should ideally be supported by the description of the uncertainty quantification in the standard methods used. The standardization of the quantification of the measurement uncertainty ensures the reporting of correct and uniform uncertainties necessary to ensure that the client is well informed regarding the measurement quality and that there is a fair competition between laboratories based on the claimed measurements quality.

7.4   Future trends

Analytical chemistry is now passing through its most important revolution, aiming at the generalization of the presentation of the (bio)chemical analytical information in an objectively interpretable way, i.e., with a clear meaning of the measured quantity and the presentation of its quality. This revolution results from the recognition of the increasing importance of analytical information to society and involves the development of concepts of measurement traceability and uncertainty in various analytical fields. The key to achieve the comparability of analytical results produced in different laboratories, and each of them with regulatory or specification limits, is traceability.

Since the publication of the ISO/IEC 17025 standard[2], the analytical community has been urged to estimate the uncertainty associated with generated analytical information. Although the accreditation bodies are having difficulties in achieving generalization of the quantification of the quality of measurements in fields that involve complex analytical methods, as in the analysis of foodstuffs, progress in this area, in test laboratories, has been considerable in the last three years. Nowadays, accredited laboratories have implemented, at least, an approximate estimation of their measurements uncertainty. Therefore, since the concept of measurement uncertainty is closely related to the concept of measurement traceability, the analytical community has been confronted with the limitations of their measurements traceability and comparability with the data produced by other laboratories. The perception of those limitations has and will catalyse the development of alternative traceability chains in various analytical fields.

In the future as the different fields of analysis realize the need to improve their measurements comparability, the metrological laboratories and the reference materials producers will be asked to produce higher quality and/or new pure chemicals and certified reference materials. The development of new analytical instrumentation will also allow the improvement of the ruggedness of the used analytical methodologies to all kinds of interferences, allowing the traceability of the information generated at the test laboratories to higher levels and, in some cases, to the SI units. For fields where traceability to the SI units is more difficult, their measurement comparability will be, at least in the next few years, supported by proficiency tests and reference methods of analysis.

The role of the Codex Alimentarius Commission (CODEX)[22] in the standardization of the characterization of foodstuffs all over the world has been extremely important, and will be more important in the future, for the comparability of measurements in this field. The CODEX, created in 1963 by the FAO (Food and Agriculture Organization of the United Nations) and the WHO (World Health Organization), aims at developing food standards, guidelines and related texts such as codes of practice under the Joint FAO/WHO Food Standards Programme. The main purposes of this programme are protecting the health of consumers, ensuring fair trade practices in the food trade, and promoting coordination of all food standards work undertaken by international governmental and non-governmental organizations. The CODEX develops its technical activities in the frame of committees, the pertaining one being the CODEX Committee on Methods of Analysis and Sampling, CCMAS.

In the near future, improvements are expected both in the methodologies used in interlaboratory evaluation of the results and in the preparation of characterized items, in this case considering the analytical equivalence of the reference items with the samples usually analysed in the laboratories[23]. Although the development of measurement comparability will be performed at different rates in the various analytical fields, even within the analysis of foodstuffs, it should be possible, in the current century, to construct internationally accepted traceability chains; not necessarily all the way up to the SI units, capable of supporting the objective comparability of measurements performed all over the world.

7.5   References

1. Valcárcel, M, Ríos, A, Maier, E, Grasserbauer, M, Nieto de Castro, C, Walsh, M C, Rius, F X, Niemelä, R, Voulgaropoulos, A, Vialle, J, Kaarls, R, Adams, F, Albus, H and Neidhart, B. 'Metrology in Chemistry and Biology: A Practical Approach', Report EUR18405 EN, European Commission, Luxemburg, 1998.
2. ISO/IEC 17025:2005, 'General requirements for the competence of testing and calibration laboratories', Geneva, Switzerland, 2005.
3. ILAC, 'The ILAC Mutual Arrangement', December 2003, www.ilac.org.
4. ISO, International vocabulary of basic and general terms in metrology (VIM), Geneva, 1993.
5. Eurachem/CITAC Guide, Traceability in Chemistry Measurement, 1st Ed., 2003, www.eurachem.ul.pt.
6. BIPM, Organization Intergovernamentale de la Convention du Mètre, Paris, 7th Edition, 1998, www.bipm.fr/en/committees/cc/ccqm.
7. Bureau International des Poids et Mesures, Proceedings of the 4th meeting of CCQM, 1998.
8. European Information System on Proficiency Testing Schemes, EPTIS, www.eptis.bam.de.
9. IRMM (Institute for Reference Materials and Measurements), EC, Geel, Belgium, www.irmm.jrc.be.
10. Quétel, C R, Nelms, S M, Van Nevel, L, Papadakis, I and Taylor, P D P. *J. Anal. At. Spectrom.*, 2001 **16**,1091–1100.
11. Aregbe, Y, Harper, C, Norgaard, J, Smet, De, M Smeyers, P, van Nevel, L and Taylor, P D P. *Accred. Qual. Assur.*, 2004, **9**, 323–332.
12. International Organization for Standardization, 'Guide to the expression of uncertainty in measurement', Geneva, Switzerland, 1995.
13. Eurachem, CITAC, 'Quantifying Uncertainty in Analytical Measurement', 2nd Ed., 2000 (www.eurachem.ul.pt).
14. Analytical Methods Committee, *Analyst*, 1995, **120**, 2303.
15. Ellison, S L R and Barwick, V J. *Accred. Qual. Assur.*, 1998, **3**, 101.
16. Ellison, S L R and Barwick, V J. *Analyst*, 1998, **123**, 1387.
17. Maroto, A, Riu, J, Boqué, R and Rius, F V. *Anal. Chim. Acta*, 1999, **391**, 173.
18. Jülicher, B, Gowik, P and Uhlig, S. *Analyst*, 1999, **124**, 537.
19. Bettencourt da Silva, R J N, Lino, M J, Santos, J R and Camões, M F G F C. *Analyst*, 2000, **125**, 1459.
20. Bettencourt da Silva, R J N, Santos, J R and Camões, M F G F C. *Analyst*, 2002, **127**, 957.
21. Bettencourt da Silva, R J N, Figueiredo, H, Santos, J R and Camões, M F G F C. *Anal. Chim. Acta*, 2003, **477**, 169.
22. Understanding The Codex Alimentarius, FAO/WHO 1999. ISBN 92-5-104248-9 www.codexalimentarius.net.
23. Belitz, H-D and Grosch, W. Food Chemistry, (Translation from German) Springer-Verlag, Berlin, 2nd edition, 1999.

Part III

Traceability technologies

8

DNA markers for animal and plant traceability

J. A. Lenstra, Utrecht University, The Netherlands

8.1 Introduction

All over the world, food production and consumption is embedded in a rich tradition with both cultural and emotional contexts. In recent decades our daily food has diversified enormously because of international trading and new technologies. At the same time, the erosion of traditional supply patterns has made the consumer more vulnerable to deliberate or accidental errors that affect the safety and quality of food. The consequences range from economic dishonesty, by which too much is paid for an inferior, adulterated or substituted product, to serious threats to consumer health by toxic, allergic or infectious agents. Recent epidemics (e.g. BSE, avian flu) as well as the regular detection in food of xenobiotics (e.g. hormones, antibiotics, weed killers and dioxins) have damaged the consumers' trust in the food industry.

These developments motivate our interest in the general issue of traceability of food, marshalling the available physical, chemical and genetic technologies for the authentication of food and identification of its geographical, biological or genetic origin. In this chapter, the contribution of modern DNA technology is described. In an earlier volume of this series (Lenstra, 2003), DNA methods for differentiation of food species were reviewed. Here, the recent progress that allows the traceability below the animal or plant species level by genetic methods is described, expanding on an earlier review by Woolfe and Primrose (2004). For the traceability of fish products, reviewed recently by Moretti *et al.* (2003), no DNA methods for differentiation below the species level have been cited yet. For the related fields of transgene detection and sexing of beef, we refer to the reviews of Zeleny and Schimmel (2002) and Auer (2003).

Since traceability depends on genetic differentiation, it is important to realize that domestic plants have a higher degree of subspecies distinctness than domestic animals. By vegetative (asexual) propagation of several crop species, the same genotype is multiplied indefinitely, giving the variety the genetic status of an immortal individual. Other domestic plants species are highly inbred by self-pollination, while, for species that can only be grown by cross-pollination, the high number of offspring from selected parents ensures that varieties or cultivars are genetically distinct. For most of the major crop species, identification of cultivars or accessions is primarily for germplasm management. However, the same methods of identification are useful for food traceability as well.

In contrast to cultivated plants, animals are propagated exclusively by sexual reproduction with only a restricted number of offspring. Despite the genetic isolation of breeds, the popularity of top sires and the systematic selection for traits of interest – morphology, production traits and automatic adaptation to the environment – differences between populations of domestic livestock are on the level of allele frequencies rather than breed-specific alleles. As a result, the development of breed traceability is not straightforward. On the other hand, it often yields information about breed relations, population genetic mechanisms and the history of the domestication process (Bruford *et al.*, 2003).

DNA analysis has a few obvious advantages over other methods of traceability: chemical stability of the DNA, a vast resource of variation at all taxonomic levels and uniform methods for the detection of any DNA sequence. Shortcomings are that most genetic markers have only an indirect relationship with phenotypic variation and that DNA sequences reveal genetic variation, but not the influence of the environment.

8.2   DNA variation at the species and subspecies level

The level of DNA variation depends on mutation rates, selective constraints and population genetic parameters such as the effective population size and selection pressure. Owing to the absence of repair mechanisms, mitochondrial DNA (mtDNA) mutates faster than nuclear DNA, resulting in an appreciable diversity within the species. The most variable segment is the control region of D-loop, which is often used for forensic purposes. However, sequence variants of mtDNA are normally not specific for breeds and the D-loop is not used for traceability below the species level.

MtDNA is suitable, however, for differentiation of species that are closely related and may be considered by food producers and consumers to be a single species. Examples are the differentiation of beef species (Verkaar *et al.*, 2002), eels (Terol *et al.*, 2002) and tuna species (Hwang *et al.*, 2004). Mitochondrial DNA is transmitted maternally, so results may be confounded by species hybridization. For this reason, complementary tests based on the species-specific nuclear satellite DNA (Verkaar *et al.*, 2002) as well as tests based on the paternally transmitted Y-chromosome (Verkaar *et al.*, 2003) have been developed for the bovine species.

Several other categories of variation in the nuclear genome have been exploited for food traceability. Per definition, any DNA sequence that is variable within species is a genetic marker. Traceability assays may target one or more markers either separately (Cunningham and Meghen, 2001) or as part of a complex electrophoretic pattern. Such a pattern is often referred to as a DNA fingerprint and does not require any information of the DNA sequences of the markers. The original DNA fingerprint stems from before the PCR (polymerase chain reaction) era and was based on the informative but time-consuming Southern blot identification of hypervariable tandemly repeated minisatellites (Jeffreys *et al.*, 1985; Fig. 8.1). Much faster is the PCR-based RAPD (random amplified polymorphic DNA; Weder, 2002; Fig. 8.2), but this technique suffers from poor reproducibility and is now considered obsolete. Comparable patterns are generated by ISSR (inter-simple sequence repeat PCR; Zietkiewicz *et al.*, 1994; Fig. 8.3) with primers specific for the tandem repeats of microsatellites. AFLP (amplified fragment length polymorphism, reviewed by Savelkoul *et al.*, 1999; Fig. 8.4) is based on a selective amplification of restriction fragments. It is now widely used in microbial and plant genetics and reproducibly displays variation that corresponds to mutations within or near the restriction sites.

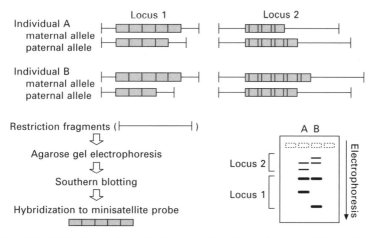

Fig. 8.1 DNA fingerprinting by Southern blotting with minisatellite probes. Restriction fragments of genomic DNA are fractionated by agarose gel electrophoresis, transferred from the gel to nylon membranes and hybridized to a probe specific for minisatellite sequences, tandemly repeated sequence motifs of 10 to 30 bp. The resulting pattern is polymorphic because of the length variation of the restriction fragments that contain a minisatellite. In this example, the repeat unit of the minisatellites in locus 2 has a few mutations relative to locus 2, which gives a less intense hybridization.

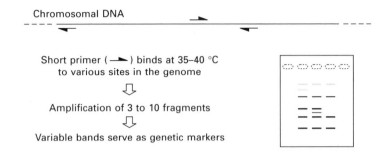

Chromosomal DNA

Short primer (⟶) binds at 35–40 °C
to various sites in the genome
⇩
Amplification of 3 to 10 fragments
⇩
Variable bands serve as genetic markers

Fig. 8.2   RAPD (random amplified polymorphic DNA). Amplification of genomic DNA with arbitrary, often short (10 bp) primers and a permissive annealing temperature of 30–40 °C generates a pattern of amplification products that appears to depend on the individual and then, per definition, provides one or more genetic markers. Most probably the RAPD polymorphism is caused by point mutations, deletions or insertions that create or abolish binding sites for the primers. Because of the low annealing temperature and the competitive amplification, the pattern may not be reproducible and often depends on the purity and concentration of the DNA template. RAPD markers are dominant, i.e., the same genotype is scored for homozygous and heterozygous presence of a band.

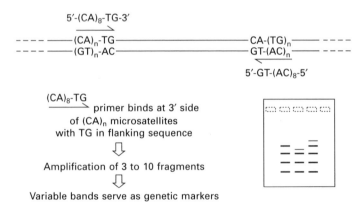

$5'$-$(CA)_8$-TG-$3'$

——— $(CA)_n$-TG ——————— CA-$(TG)_n$ ———
——— $(GT)_n$-AC ——————— GT-$(AC)_n$ ———

$5'$-GT-$(AC)_8$-$5'$

$(CA)_8$-TG
⟶ primer binds at 3′ side
of $(CA)_n$ microsatellites
with TG in flanking sequence
⇩
Amplification of 3 to 10 fragments
⇩
Variable bands serve as genetic markers

Fig. 8.3   ISSR (Inter simple-sequence repeat) PCR. In this figure, both strands of the genomic DNA are indicated. PCR primers are specific for microsatellite sequences like $(CA)_n$ and induce the amplification of DNA segments between microsatellites. Shown is a primer with a 3′ extension TG that binds to AC in the flanking sequence. This reduces the number of amplified fragments, but also anchors the primer to the 3′ end of the microsatellite. Alternatively, a 5′ extension may direct the primer to the 5′ end.

There is now a clear trend, however, towards dedicated assays for genotyping of separate genetic markers. Several reports use the SCAR (sequence characterized amplified region) approach: identification of anonymous breed-specific DNA sequences by sequencing of RAPD (random-amplified polymorphic DNA) fragments with the desired specificity, followed by the design of PCR primers for their specific detection. However, any PCR assay

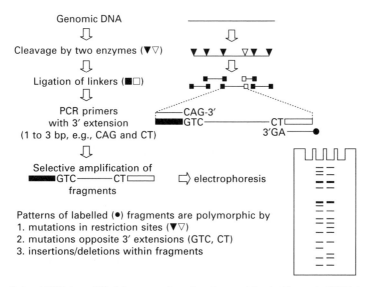

Fig. 8.4 AFLP (amplified fragment length polymorphism). Genomic DNA is cut with a combination of enzymes and ligated to linkers. PCR is carried out with primers binding to the linker and the restriction site with an extension of 1 to 3 nucleotides. Depending on genome size and the length of the extension, PCR leads to the selective amplification on 50 to 100 restriction fragments that have next to the restriction site nucleotides complementary to the extension. Separation on a sequencing gel then generates a pattern in which presence/absence polymorphisms correspond to mutations in the restriction site, to mutations in the sequence complementary to the extension or to deletions/insertions within the fragment.

that is based on the amplification/no amplification of a fragment is prone to giving false positives because of contamination.

Much more robust is the amplification of a marker with two or more alleles that reveal the origin or identity of the sample. As exemplified below, most recently developed methods are based on the reproducible and informative microsatellites (Fig. 8.5), often also denoted as SSRs for simple-sequence repeat, as STR for short tandem repeat or as SSP for simple-sequence polymorphism.

8.3   Traceability below the species level

8.3.1   Livestock

Meat and dairy products from livestock species are the central components of our meals and account for a substantial part of our protein uptake. A relatively high risk of microbial and often zoonotic infections emphasizes the requirement for traceability tests. Furthermore, the perceived, if not real, quality of meat and cheese is often linked to their origin.

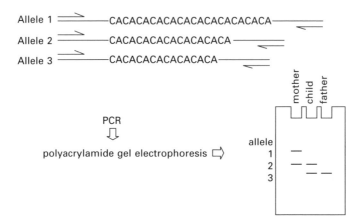

Fig. 8.5 Microsatellites are tandem repeats with a short repeat unit of 2, 3 or 4 bp. In mammals, the most common is the repeat of CACACA—CACA, which occurs about 100 000 times in a mammalian genome. Several microsatellites are polymorphic and have up to 10 or more different alleles and a low degree of homozygozity. PCR with primers specific for the flanking sequences and length separation on a sequencing gel generate for heterozygous individuals two different alleles, one inherited from the mother and the other from the father.

The use of microsatellites or other neutral genetic markers for breed identification is complicated by the variability within breeds (Hall, 2004). Furthermore, the reported success rate depends entirely on the test panel of breeds, each of which are subject to stochastic and temporal fluctuations of allele frequencies. However, modest successes with breed assignments on the basis of microsatellites have been reported: 62% correct assignment with 19 microsatellites in seven north-western European cattle breeds (Manel *et al.*, 2002) and 88 to 100% with 20 microsatellites in a panel of Holstein-Friesian and four Italian white cattle breeds (Ciampolini *et al.*, 2000).

In contrast, the use of microsatellites for individual identification (Arana *et al.*, 2002; Vázquez *et al.*, 2004) is straightforward with a matching probability for random animals of 0.001 with only three markers. However, microsatellite genotyping requires a rigorous standardization with allelic ladders or representative standard animals for a comparison of results across laboratories. Two reports independently developed SNP assays for identification and paternity testing with 32 SNPs (single-nucleotide polymorphisms) validated for US beef cattle (Heaton *et al.*, 2002) and 37 SNPs validated for Holstein, Simmental and Swiss Brown dairy cattle (Werner *et al.*, 2004).

Sasazaki *et al.* (2004) demonstrated the use of AFLP for selection of fragments specific for Japanese Black cattle. Similarly, Fumière *et al.* (2003) found strain-specific AFLP markers in chicken. Since the time-consuming conversion of AFLP polymorphisms to SNPs (Sasazaki *et al.*, 2004) has become more feasible by the recent availability of the complete genomic sequences, AFLP may allow a practical way for the selection of markers with a desired specificity.

For several cattle or porcine breeds, specific markers may be derived from genes encoding the coat colour (Olson, 1999; Kijas *et al.*, 1998, 2001; Klungland and Vage, 2003). A deletion in the gene-encoding melanocortin receptor 1 in the extension locus (MC1-R or MSHR) was found to cause the red colour in Holstein cattle (Joerg *et al.*, 1996). Rouzaud *et al.* (2000) obtained a partial differentiation of eight French cattle breeds, on the basis of four MC1-R alleles. One of these alleles was suitable for detecting Holstein milk in French cheese for which the RDO (registered designation of origin) stamp imposes restrictions on the origin of the cattle breed (Maudet and Taberlet, 2002).

Kriegesmann *et al.* (2001) identified two other breed-specific MC1-R alleles in Salers and Swiss-Brown cattle, respectively, but these alleles were not fixed in these breeds.

Kijas *et al.* (1998) described four MC1-R alleles that were fixed in several pig breeds. The specificity of the MC1R*4 allele for the Duroc breed was used for the detection of this breed in Iberian meat products (Fernández *et al.*, 2004).

A special case of the DNA-based traceability test is the analysis of bacterial 16S rRNA composition in natural whey cultures for the manufacture of water-buffalo mozzarella, which was found to correlate with the geographical origin (Mauriello *et al.*, 2003).

8.3.2   Cereals

Along with potatoes and cassava, cereals are the major source of calories. Prepared by baking or boiling, cereal dishes provide carbohydrate in our diet and substance in our meals.

Rice (*Oryza sativa*) is the major energy source for more than half of the world population. Cultivars differ in their aromatic content and can be differentiated by RAPD (Choudhury *et al.*, 2001), but more reliably by AFLP (Bligh *et al.*, 1999) or microsatellite typing. Garland *et al.* (1999) used 10 microsatellites for identification of Australian breeding lines, while Nagaraju *et al.* (2002) identified microsatellite and ISSR alleles that distinguished 24 traditional Basmati, evolved Basmati and semidwarf non-Basmati varieties.

Wheat (*Triticum aestivum*) is now the main component of bread. Thanks to the elasticity of the gluten proteins, it has superior baking qualities. Detection of glutenin alleles in a multiplex PCR has been proposed as a tool for identification of wheat varieties (Ma *et al.*, 2003). The best quality gluten is found in durum varieties (*Triticum turgidum*, var. *durum*), which are used in the manufacture of pasta. For cultivar differentiation, several microsatellite markers have been characterized (Pasqualone *et al.*, 1999; Prasad *et al.*, 2000; Dograr *et al.*, 2000). Perry (2004) developed a multiplex PCR of seven microsatellites to differentiate Canadian durum varieties. Röder *et al.* (2002) used a different set of 10 microsatellites for the construction of a database of European wheat varieties. However, with a largely overlapping set of nine markers, Cooke *et al.* (2004) found that only 24 out of 45 varieties of wheat met the standards of distinctness, uniformity and stability (DUS).

Barley (*Hordeum vulgare*) is used in beer production and for fodder. Differentiation of the many available varieties has been accomplished by RAPD and ISSR (Fernández *et al.*, 2002). A systematic study of the variability of 65 microsatellites (Sjakste *et al.*, 2003) showed that 14 out of 36 Latvian varieties and related European ancestors were completely homogeneous.

Because of the lack of gluten, flour from maize (*Zea mays*) has poor raising capabilities. Its main use is in feed for livestock, but in different forms it is also used for human consumption. Modern production is based on the development and hybridization of inbred lines. Gethi *et al.* (2002) found by analysis with 44 microsatellites that 88% of the variation is across inbred lines, but also that a significant variation exists among sources of the same line or within lines.

Sorghum [*Surhum bicolor* (L.) Moench] is related to maize and 80% of it is grown in Africa and Asia, where it tolerates arid growth conditions. Microsatellite analysis (Djè *et al.*, 2000) revealed only a limited degree of distinctiveness of different accessions with only 70% of the variation among accessions.

The related pearl millet (*Pennietum glaucum*) belongs with rice, wheat, maize, barley and sorghum in the most important food crops group. Like sorghum, it can be grown in arid areas. The different landraces offer a variety of phenotypes, which can be genotyped by RAPD or Southern blotting with a $(GATA)_4$ probe (Chowdari *et al.*, 1998).

8.3.3   Vegetables

Vegetables accompany every dinner as supplementary sources of energy, proteins, fats, minerals and/or water-soluble vitamins. The potato (*Solanum tuberosum*) is one of the main suppliers of energy in the temperate regions. Corbett *et al.* (2001) found that microsatellites were more reliable than ISSR and that the 50 common UK varieties could be differentiated with only three markers. For the documentation of the genetic identity of ex situ potato cultivars, Ghislain *et al.* (2004) developed a set of 18 informative markers that are applicable to all eight cultivar groups.

Carrots (*Daucus carota* spp. s*ativus*) are also a major vegetable crop and an essential source of provitamin A. It is an outcrossing species, for which AFLP patterns indicated that most of the genetic variation is within rather than between lines (Shim and Jørgensen, 2000). The evaluation of seed purity requires a differentiation of parental lines and F_1 hybrids. Although AFLP on pooled samples revealed line-specific fragments, parental lines are too heterogeneous for testing hybrid seed purity (Grzebelus *et al.*, 2001).

The common bean (*Phaseolus vulgaris* L.) is an important source of protein. Métais *et al.* (2002) developed microsatellite markers and analysed the genetic diversity in various bean types. Galván *et al.* (2003) used ISSR markers in order to identify the South-American gene pool of origin.

Soybean [*Glycine max* (L.) Merr.] is a dominant source of protein and oil in both human food and animal feeds. For the purpose of Plant Variety Protection (PVP), cultivars that were identical for morphological and pigmentation traits were distinguished by typing 20 microsatellites (Diwan and Cregan, 1997).

Rapeseed (*Brassica napus*) is grown in temperate and cool climate zones and is also important as oil and fodder crop. Hellebrand *et al.* (1998) developed procedures for isolation of DNA in vegetable oil. Sobotka *et al.* (2004) found that only 8 out of 69 RAPD primers and 5 out of 15 microsatellite primer pairs generated polymorphic bands and that AFLP patterns were the most informative for cultivar identification. Plieske and Struss (2001) described more polymorphic microsatellites for rapeseed, several of which (83%) could also be amplified in other *Brassica* species. Tommasini *et al.* (2003) developed three multiplex sets of five microsatellite primer pairs and successfully assigned 99% of the plants to the correct variety. ISSR patterns of pooled DNA samples differentiated cultivars from rapeseed and from the related turnip (*Brassica rapa*), but patterns from individuals were too heterogeneous for identification of individuals (Charters *et al.*, 1996).

Zheng *et al.* (2001) reported that AFLP is superior to isozymes and RAPD for the identification of Chinese cabbage (*Brassica campestris* L. ssp. Pekinensis) cultivars.

8.3.4 Fruits

The seductive power of fruits from a large variety of species is an adaptation to their role in the dispersal of seeds and has been refined by cultivation. Fruit has also long been a symbol of food abundance and good health.

Wünsch and Hormaza (2002a) already reviewed DNA-based cultivar identification of temperate fruit trees, including the use of microsatellites for apple (*Malus x domestica*), European (*Pyrus communis*) and Japanese (*Pyrus pyrifolia*) pear, peach (*Prunus persica*), apricot (*Prunus armeniaca*), sweet cherry (*Prunus avium*, Wünsch and Hormaza, 2002b), black cherry (*Prunus serotina*) and olives (*Olea europaea*). Microsatellite primers can often be used for related species: primers developed for apple were useful in another important pome fruit, pear, while several primers for the stone fruit peach can be used for other *Prunus* species (see also Serrano *et al.*, 2002; Dirlewanger *et al.*, 2002).

A more recent PCR-RFLP analysis of the 18S rDNA gene in European and Japanese pear (Lee *et al.*, 2004) only differentiated the cultivars incompletely.

Peach is, after apple, the second most important fruit crop in the temperate and subtropical zones. The differentiation of peach and nectarine cultivars was worked out further by Aranzana *et al.* (2003). Cultivars of apricot, another economically important stone fruit, differ considerably in morphology and growth properties. Using different sets of 14 microsatellite primers derived from peach, two groups demonstrated discrimination of cultivars. (Zhebentyayeva *et al.*, 2003; Sánchez-Pérez *et al.*, 2005). Typing almonds (*Prunus dulcis*) with peach-derived primers revealed a considerable diversity among cultivars with more alleles than found in peaches (Martínez-Goméz *et al.*, 2003).

Citrus fruits (*Citrus* spp.) are the most important fruits from tropical and subtropical areas with orange (*Citrus sinensis*) accounting for more than half of the production. Genetic identification of rootstock cultivars and differentiation of zygotic and asexual seedlings are most essential for breeding. Ruiz *et al.* (2000) developed a microsatellite-based identification of sexual seedlings. Abkenar and Isshiki (2003) showed that RAPD allowed a distinction of closely related Japanese cultivars.

The use of microsatellites was also extended to the tropical lychee (*Litchi chinensis*) by isolation of 12 microsatellites (Viruel and Hormaza, 2004), four of which were sufficient to differentiate 16 cultivars as well as four cultivars of the related longan *(Euphoria longan)*.

Kiwifruit (*Actinidia deliciosa* A. Chev) is a recently domesticated plant. For this outcrossing species, female and male cultivars are propagated in vitro, which may induce somaclonal variation. For checking the genetic integrity of the parent cultivars, microsatellites were more informative than RAPD (Palombi and Damiano, 2002).

The vegetative propagation of cultivated strawberry (*Fragaria × ananassa*) requires a reliable check of the starting material. For this, ISSR (Arnau *et al.*, 2002) and AFLP (Tyrka *et al.*, 2002) were found to be better reproducible than RAPD (Degani *et al.*, 1998). Microsatellite primers developed for the wild diploid relative *Fragaria vesca* (Hadonou *et al.*, 2004) as well as 12 out of 41 of *Prunus* specific primers (Dirlewanger *et al.*, 2002) also worked for the cultivated strawberry and will probably be useful for cultivar identification.

Grape (*Vitis vinifera*) derives its importance almost solely from wine production. There is no other crop food commodity for which the values depend, as absolutely as for wine, on subtle differences in fragrance and flavour. For this reason, a wine depends heavily on the cultivar of the grape and authenticity tests are essential. Cultivar identification has been described on the basis of AFLP (Cervera *et al.*, 1998) and, with limited success, on the basis of SCAR primers (Vidal *et al.*, 2000). García-Beneytez *et al* (2002) used a microsatellite multiplex PCR works for monovarietal musts, but this was not successful for DNA extracts from wine. With an overlapping set of six microsatellites, This *et al.* (2004) standardized the identification of cultivars. However, another overlapping set of microsatellites revealed total identity for 24 Traminer grapevine accessions for which AFLP detected 16 different genotypes (Imazio *et al.*, 2002). Interestingly, methylation-sensitive AFLP changed for two accessions the genetic distances to the other accessions, suggesting a role for epigenetic modifications.

Because of their lack of sweetness, tomatoes (*Lycipersicon esculentum*) as well as olives (*Olea europaea*) join the legume dishes rather than being consumed as healthy snacks. There are large differences between cultivars of tomato with respect to season maturity, colour, shape, size and taste. Four microsatellites from a selected set of 20 markers were sufficient for discrimination of all 16 cultivars tested (Bredemeijer *et al.*, 1998). Bredemeijer *et al.* (2002) published a database of profiles of 20 microsatellites for more than 500 varieties. Subsequently (Cooke *et al.*, 2004), six of these loci were used to demonstrate that 9 out of 10 varieties were distinct, uniform and stable (DUS). A separate group developed other microsatellite markers and analysed relationships between cultivars (He *et al.*, 2003).

For the vegetatively propagated olive tree, cultivars are linked to the region of origin and their identification is most relevant for quality assurance. DNA extraction from oil (Busconi *et al.*, 2003; Breton *et al.*, 2004) allows the identification of cultivars in the end product. After the development of SCAR (Hernández *et al.*, 2001; Bautista *et al.*, 2002) and ISSR (Pasqualone *et al.*, 2001) assays, the most recent studies focus on microsatellites (Cipriani *et al.*, 2002; Pasqualone *et al.*, 2004) and showed that three markers differentiate 10 Italian strains.

8.3.5 Other plants

Several plants are grown not because of any nutritional value, but for their taste, scent or psychic effects. Their products are used in beverages, as spice, as medicine or for inhalation. For identification of cultivars of several of these species DNA tests are available.

Tea, coffee and hot chocolate are of Chinese, Arabian and Aztec origin, respectively, and are now the most important hot beverages worldwide. Varieties of green tea [*Camellia sinensis* (L.) O. Kuntzem var. *Sinensis*] were differentiated by STS-RFLP analysis of nuclear genes (Kaundun and Matsumoto, 2003). Microsatellites for tea have been developed (Freeman *et al.*, 2004) and are also informative for cultivar identification (http://www.niab.com/FILEAREA/pos0008zh.pdf). The diversity of coffee (*Coffea arabica*) cultivars has been investigated with AFLP and microsatellite markers (Anthony *et al.*, 2002). Microsatellite markers (Baruah *et al.*, 2003) differentiated beans from different origins (http://www.niab.com/FILEAREA/pos0008zh.pdf). As for several other plant species, the genetic analysis of cocoa (*Theobroma cacao*) moved from RAPD to AFLP (Perry *et al.*, 1998) and then to microsatellites (Saunders *et al.*, 2000).

Ginseng (*Panax ginseng*) and American ginseng (*Panax quinguefolius*) are sources of traditional Chinese medicine. These species can be differentiated by AFLP and minisatellite typing (Ha *et al.*, 2002). Hon *et al.* (2003) reviewed DNA methods for genetic authentication of both ginseng species and achieved, with microsatellite markers, a resolution down to the farm level. Alternatively, Mihalov *et al.* (2000) described a PCR on the nuclear ribosomal internal transcribed spacer and the chloroplast ribulose 1,5-biphosphate carboxylase large subunit for differentiation of commercial samples.

Chili pepper (*Capsicum annuum*) is cultivated as a hybrid. Jang *et al.* (2004) used RAPD and SCAR markers for purity testing of the F1 seed, which may be contaminated by self-pollination of the parents.

Hemp or marijuana (*Cannabis sativa*) is well known as a source of intoxicant, but it also provides fibre for ropes, feeds, oils and medicine. Since trading and using the psychoactive products is prohibited in most countries, traceability is directly relevant for forensic purposes. Miller Coyle *et al.* (2003) reviewed different methods for identification and validated an AFLP-based typing. Microsatellite typing has been developed by Alghanim and Almirall (2003) and by Gilmore and Peakall (2003).

Despite an increasing awareness of negative health effects, tobacco is the world's most commonly occurring psychosocial addiction. AFLP analysis revealed only a limited amount of polymorphism among cultivars (Ren and Timko, 2001).

8.4   Future trends

In recent years, there has been a clear trend, first from RAPD and ISSR towards SCAR and AFLP and then towards typing of microsatellites. Their reproducibility and informativeness with regard to distinctness, relationships between breeds and traceability outweighs the initial investment in marker isolation and primer development.

Cultivars of most major crop species can now be differentiated by microsatellite typing. For animals, diversity within breeds is a confounding factor, which can be overcome only by using several (typically 20 to 30) markers. A better differentiation of breeds may be accomplished by using gene variants that have a causative link to breed-specific phenotypes as the coat colours of cattle and swine (see above). Complete genomic sequences of several species, the accompanying collections of SNPs, efficient technologies for SNP genotyping and the growing amount of information about gene function now promise that this will be a future trend.

SNPs are generally nominated as successors of the microsatellites as neutral genetic markers for mapping studies with five SNPs having about the same power as one microsatellite. SNPs are more abundant than microsatellites and can be identified by alignment of genomic or mRNA sequences from different sources, often generated during whole-genome sequencing. However, their general use for identification and traceability will depend on the availability of practical and cost-effective methods for SNP typing that are tailored to the low- or medium throughput required for food analysis.

8.5 Acknowledgement

We acknowledge the support of the Framework 6 project TRACE (Tracing Food Commodities in Europe) funded by the European Union. The content of this publication does not represent the views of the Commission or its services.

8.6 References

Abkenar, A and Isshiki, S 2003, Molecular characterization and genetic diversity among Japanese acid citrus (*Citrus* spp.) based on RAPD markers, *J. Hortic. Sci. Biotechnol.* **78**, 104–107.

Alghanim, H J and Almirall, J R 2003, Development of microsatellite markers in *Cannabis sativa* for DNA typing and genetic relatedness analyses, *Anal. Bioanal. Chem.*, **376**, 1225–1233.

Anthony, F, Combes, C, Astorga, C, Bertrand, B, Graziosi, G and Lashermes, P 2002, The origin of cultivated *Coffea arabica* L. varieties revealed by AFLP and SSR markers, *Theor. Appl. Genet.*, **104**, 894–900.

Arana, A, Soret, B, Lasa, I and Alfonso, L 2002, Meat traceability using DNA markers: application to the beef industry, *Meat Sci.*, **61**, 367–373.

Aranzana, M J, Carbo, J and Arus, P 2003, Microsatellite variability in peach [*Prunus persica* (L.) Batsch]: cultivar identification, marker mutation, pedigree inferences and population structure, *Theor. Appl. Genet.*, **106**, 1341–1352.

Arnau, G, Lallemand, J and Bourgoin, M 2002, Fast and reliable strawberry cultivar identification using inter simple sequence repeat (ISSR) amplification, *Euphytica*, **129**, 69–79.

Auer, C A 2003, Tracking genes from seed to supermarket: techniques and trends, *Trends Plant Sci.*, **8**, 591–597.

Baruah, A, Naik, V, Hendre, P S, Rajkumar, R, Rajendrakumar, P and Aggarwal, R K 2003, Isolation and characterization of nine microsatellite markers from *Coffea arabica* L., showing wide cross-species amplifications, *Molec. Ecol. Notes.*, **3**, 647–650.

Bautista, R, Crespillo, R, Canovas, F M and Claros, M G 2002, Identification of olive-tree cultivars with SCAR markers', *Euphytica*, **129**, 33–41.

Bligh, H F J, Blackhall, N W, Edwards, K J and McClung, A M 1999, Using amplified fragment length polymorphisms and simple sequence length polymorphisms to identify cultivars of brown and white milled rice, *Crop Sci.*, **39**, 1715–1721.

Bredemeijer, G M M, Arens, P, Wouters, D, Visser, D, and Vosman, B 1998, The use of semi-automated fluorescent microsatellite analysis for tomato cultivar identification, *Theor. Appl. Genet.*, **97**, 584–590.

Bredemeijer, M, Cooke, J, Ganal, W, Peeters, R, Isaac, P, Noordijk, Y, Rendell, S, Jackson, J, Roder, S, Wendehake, K, Dijcks, M, Amelaine, M, Wickaert, V, Bertrand, L and Vosman, B 2002, Construction and testing of a microsatellite database containing more than 500 tomato varieties, *Theor. Appl. Genet.*, **105**, 1019–1026.

Breton, C, Claux, D, Metton, I, Skorski, G and Berville, A 2004, Comparative study of methods for DNA preparation from olive oil samples to identify cultivar SSR alleles in commercial oil samples: possible forensic applications, *J. Agric. Food. Chem.*, **52**, 531–537.

Bruford, M W, Bradley, D G and Luikart, G 2003, DNA markers reveal the complexity of livestock domestication, *Nat. Rev. Genet.*, **4**, 900–910.

Busconi, M, Foroni, C, Corradi, M, Bongiorni, C, Cattapan, F and Fogher, C 2003, DNA extraction from olive oil and its use in the identification of the production cultivar, *Food Chem.*, **93**, 127–134.

Cervera, M-T, Cabezas, J A, Sancha, J C, Martínez, D E, Toda, F and Martínez-Zapater, J M 1998, Application of AFLPs to the characterization of grapevine *Vitis vinifera* L. genetic resources. A case study with accessions from Rioja (Spain), *Theor. Appl. Genet.*, **97**, 51–59.

Charters, Y M, Robertson, A, Wilkinson, M J and Ramsay, G 1996, PCR analysis of oilseed rape cultivars (*Brassica napus* L. ssp. Oleifera) using 5′-anchored simple sequence repeat (SSR) primers, *Theor. Appl. Genet.*, **92**, 442–447.

Choudhury, P R, Kohli, S, Srinivasan, K, Mohapatra, T and Sharma, R P 2001, Identification and classification of aromatic rices based on DNA fingerprinting, *Euphytica*, **118**, 243–251.

Chowdari, K V, Davierwala, A P, Gupta, V S, Ranjekar, P K and Govila, O P 1998, Genotype identification and assessment of genetic relationships in pearl millet [*Pennisetum glaucum* (L.) R. Br] using microsatellites and RAPDs, *Theor. Appl. Genet.*, **97**, 154–162.

Ciampolini, R, Leveziel, H, Mazzanti, E, Grohs, C and Cianci, D 2000, Genomic identification of the breed of an individual or its tissue, *Meat Sci.*, **54**, 335–340.

Cipriani, G, Marrazzo, M T, Marconi, R, Cimato, A and Testolin, R 2002, Microsatellite markers isolated in olive (*Olea europaea* L.) are suitable for individual fingerprinting and reveal polymorphism within ancient cultivars, *Theor. Appl. Genet.*, **104**, 223–228.

Cooke, R J, Bredemeijer, G M M, Ganal, G W, Peeters, R, Isaac, P, Rendell, S, Jackson, J, Röder, M S, Korzun, V, Wendehake, K, Areshchenkova, T, Dijcks, M, Laborie, D, Bertrand, L and Vosman, B 2004, Assessment of the uniformity of wheat and tomato varieties at DNA microsatellite loci, *Euphytica*, **132**, 331–341.

Corbett, G, Lee, D, Donini, P and Cooke, R J 2001, Identification of potato varieties by DNA profiling, *Acta Horticulturae*, **546**, 387–390.

Cunningham, E P and Meghen, C M 2001, Biological identification systems: genetic markers, *Rev. Sci. Tech. Off. Int. Epiz.*, **20**, 491–499.

Degani, C, Rowland, L J, Saunders, J A, Hokanson, S C, Hortynski, E L and Galletta, G J 1998, The use of random amplified polymorphic DNA (RAPD) markers to identify strawberry varieties: and advanced breeding lines, *Euphytica*, **102**, 247–253.

Dirlewanger, E, Cosson, P, Tavaud, M, Aranzana, J, Poizat, C, Zanetto, A, Arus, P and Laigret, F 2002, Development of microsatellite markers in peach [*Prunus persica* (L.) Batsch] and their use in genetic diversity analysis in peach and sweet cherry (*Prunus avium* L.), *Theor. Appl. Genet.*, **105**, 127–138.

Diwan, N and Cregan, P B 1997, Automated sizing of fluorescent labeled simple sequence repeat markers to assay genetic variation in soybean, *Theor. Appl. Genet.*, **97**, 723–733.

Djè, Y, Heuertz, M, Lefèbvre, C and Vekemans, X 2000, Assessment of genetic diversity within and among germplasm accessions in cultivated sorghum using microsatellite markers, *Theor. Appl. Genet.*, **100**, 918–925.

Dograr, N, Akin-Yalin, S and Akkaya, M S 2000, Discriminating durum wheat cultivars using highly polymorphic simple sequence repeat DNA markers, *Plant Breeding*, **119**, 360–362.

Fernández, A, Fabuel, E, Alves, E, Rodriguez, C, Silió, L and Óvilo, C 2004, DNA tests based on coat colour genes for authentication of the raw material of meat products from Iberian pigs, *J. Sci. Food Agric.*, **84**, 1855–1860.

Fernández, E, Figueiras, M and Benito, C 2002, The use of ISSR and RAPD markers for detecting DNA polymorphism, genotype identification and genetic diversity among barley cultivars with known origin, *Theor. Appl. Genet.*, **104**, 845–851.

Freeman, S, West, J, James, C, Lea, V and Mayes, S 2004, Isolation and characterization of highly polymorphic microsatellites in tea (*Camellia sinensis*), *Mol. Ecol. Notes*, **4**, 324–326.

Fumière, O, Dubois, M, Gregoire, D, Thewis, A and Berben, G, 2003, Identification on commercialized products of AFLP markers able to discriminate slow-from fast-growing chicken strains, *J. Agric. Food Chem.*, **51**(5), 1115–1119.

Galván, M Z, Bornet, B, Balatt, P A and Branchard, M, 2003, Inter simple sequence repeat (ISSR) markers as a tool for the assessment of both genetic diversity and gene pool origin in common bean (*Phaseolus vulgaris* L.), *Euphytica*, **132**, 297–301.

Garcia-Beneytez, E, Moreno-Arribas, M V, Borrego, J, Polo, M C and Ibanez, J 2002, Application of a DNA analysis method for the cultivar identification of grape musts and experimental and commercial wines of *Vitis vinifera* L. using microsatellite markers, *J. Agric. Food Chem.*, **50**, 6090–6096.

Garland, S H, Lewin, L, Abedinia, M, Henry, R and Blakeney, A 1999, The use of microsatellite polymorphisms for the identification of Australian breeding lines of rice (*Oryza sativa* L.), *Euphytica*, **108**, 53–63.

Gethi, J G, Labate, J A, Lamkey, K R, Smith, M E and Kresovich, S 2002, SSR variation in important U.S. maize inbred lines, *Crop Sci.*, **42**, 951–957.

Ghislain, M, Spooner, D M, Rodriguez, F, Villamon, F, Nunez, J, Vasquez, C, Waugh, R and Bonierbale, M 2004, Selection of highly informative and user-friendly microsatellites (SSRs) for genotyping of cultivated potato, *Theor. Appl. Genet.*, **108**, 881–890.

Gilmore, S and Peakall, R 2003, Isolation of microsatellite markers in *Cannabis sativa* L. (marijuana), *Mol. Ecol. Notes*, **3**, 105–107.

Grzebelus, D, Senalik, D, Jagosz, B, Simon, P W And Michalik, B 2001, The use of AFLP markers for the identification of carrot breeding lines and F1 hybrids, *Plant Breeding*, **120**, 526–528.

Ha, W Y, Shaw, P C, Liu, J, Yau, F C and Wang, J 2002, Authentication of *Panax ginseng* and *Panax quinquefolius* using amplified fragment length polymorphism (AFLP) and directed amplification of minisatellite region DNA (DAMD), *J. Agric. Food Chem.*, **50**, 1871–1875.

Hadonou, A M, Sargent, D J, Wilson, F, James, C M and Simpson, D W 2004, Development of microsatellite markers in Fragaria, their use in genetic diversity analysis, and their potential for genetic linkage mapping, *Genome*, **47**, 429–438.

Hall, S J G 2004, Livestock biodiversity. Genetic resources for the farming of the future, Blackwell, Oxford.

He, C, Poysa, V and Yu, K 2003, Development and characterization of simple sequence repeat (SSR) markers and their use in determining relationships among *Lycopersicon esculentum* cultivars, *Theor. Appl. Genet.*, **106**, 363–373.

Heaton, M P, Harhay, G P, Bennett, G L, Stone, R T, Grosse, W M, Casas, E, Keele, J W, Smith, T P, Chitko-Mckown, C G and Laegreid, W W 2002, Selection and use of SNP markers for animal identification and paternity analysis in U.S. beef cattle, *Mammal Genome*, **13**, 272–281.

Hellebrand, M, Nagy, M and Mörsel, J-T 1998, Determination of DNA traces in rapeseed oil, *Z. Lebensm. Unters. Forsch. A*, **206**, 237–242.

Hernández, P, De, L A, Rosa, R, Rallo, L, Dorado, G and Martín, A 2001, Development of SCAR markers in olive (*Olea europaea*) by direct sequencing of RAPD products: applications in olive germplasm evaluation and mapping, *Theor. Appl. Genet.*, **103**, 788–791.

Hon, C C, Chow, Y C, Zeng, F Y and Leung, F C 2003, Genetic authentication of ginseng and other traditional Chinese medicine, *Acta Pharmacol. Sin.*, **24**, 841–846.

Hwang, D F, Jen, H C, Hsieh, Y W and Shiau, C Y 2004, Applying DNA techniques to the identification of the species of dressed toasted eel products, *J. Agric. Food Chem.*, **52**, 5972–5977.

Imazio, S, Labra, M, Grassi, F, Winfield, M, Bardini, A and Scienza, A 2002, Molecular tools for clone identification: the case of the grapevine cultivar 'Traminer', *Plant Breeding*, **121**, 531–535.

Jang, I, Moon, J H, Yoon, J B, Yoo, J H, Yang, T J, Kim, Y J and Park, H G 2004, Application of RAPD and SCAR markers for purity testing of F1 hybrid seed in chili pepper (*Capsicum annuum*), *Mol. Cells*, **18**, 295–299.

Jeffreys, A J, Wilson, V and Thein, S 1 1985, Hypervariable 'minisatellite' regions on human DNA, *Nature*, **314**, 67–73.

Joerg, H, Fries, H R, Meijerink, E and Stranzinger, G F 1996, Red coat color in Holstein cattle is associated with a deletion in the MSHR gene, *Mammal Genome*, **7**, 317–318.

Kaundun, S S and Matsumoto, S 2003, Identification of processed Japanese green tea based on polymorphisms generated by STS-RFLP analysis, *J. Agric. Food Chem.*, **26**, 1765–1770.

Kijas, J H M, Wales, R, Törnsten, A, Chardon, P, Moller, M and Andersson, L 1998, Melanocortin receptor 1 (*MC1R*) mutations and coat color in pigs, *Genetics*, **150**, 1177–1185.

Kijas, J M H, Moller, M, Plastow, G and Andersson, L 2001, A frameshift mutation in *MC1R* and a high frequency of somatic reversions cause black spotting in pigs, *Genetics*, **158**, 779–85.

Klungland, H and Vage, D I 2003, Pigmentary switches in domestic animal species, *Ann. N. Y. Acad. Sci.*, **994**, 331–338.

Kriegesmann, B, Dierkes, B, Leeb, T, Jansen, S and Brenig, B 2001, Two breed-specific bovine *MC1-R* alleles in Brown Swiss and Saler breeds, *J. Dairy. Sci.*, **84**, 1768–1771.

Lee, G P, Lee, C H and Kim, C S 2004, Molecular markers derived from RAPD, SCAR, and the conserved 18S rDNA sequences for classification and identification in *Pyrus pyrofolia* and *P. communis*, *Theor. Appl. Genet.*, **108**, 1487–1491.

Lenstra J A, 2003, DNA methods for identifying plant and animals species in foods, Chapter 3 in *Food authenticity and traceability*, Woodhead Publishing, Cambridge, UK, pp. 34–53.

Ma, W, Zhang, W and Gale, K R 2003, Multiplex-PCR typing of high molecular weight glutenin alleles in wheat, *Euphytica*, **134**, 51–60.

Manel, S, Berthier, P and Luikart, G 2002, Detecting wildlife poaching: identifying the origin of individuals with Bayesian assignment tests and multilocus genotypes, *Conserv. Biol.*, **16**, 650–659.

Martínez-Gómez, P., Arulsekar, S., Potter, D and Gradziel, T M 2003, An extended interspecific gene pool available to peach and almond breeding as characterized using simple sequence repeat (SSR) markers, *Euphytica*, **131**, 313–322.

Maudet, C and Taberlet, P 2002, Holstein's milk detection in cheeses inferred from melanocortin receptor 1 (*MC1R*) gene polymorphism, *J. Dairy Sci.*, **85**, 707–715.

Mauriello, G, Moio, L, Genovese, A and Ercolini, D 2003, Relationships between flavoring capabilities, bacterial composition, and geographical origin of natural whey cultures used for traditional water-buffalo mozzarella cheese manufacture, *J. Dairy Sci.*, **86**, 486–497.

Métais, I, Hamon, B, Jalouzot, R and Peltier, D 2002, Structure and level of genetic diversity in various been types evidenced with microsatellite markers isolated from a genomic enriched library, *Theor. Appl. Genet.*, **104**(8), 1346–1352.

Mihalov, J J, Marderosian, A D and Pierce, J C 2000, DNA identification of commercial ginseng samples, *J. Agric. Food Chem.*, **48**, 3744–3752.

Miller Coyle, H, Palmbach, T, Juliano, N, Ladd, C and Lee, H C 2003, An overview of DNA methods for the identification and individualization of marijuana, *Croat. Med. J.*, **44**, 315–321.

Moretti, V M, Turchini, G M, Bellagamba, F and Caprino, F 2003, Traceability issues in fishery and aquaculture products, *Vet. Res. Commun.*, **27** Suppl 1, 497–505.

Nagaraju, J, Kathirvel, M, Kumar, R R, Siddiq, E A and Hasnain, S E, 2002, Genetic analysis of traditional and evolved Basmati and non-Basmati rice varieties by using fluorescence-based ISSR-PCR and SSR markers, *Proc. Natl. Acad. Sci. USA*, **99**, 5836–5841, 13357.

Olson, T A 1999, In *The Genetics of Cattle* (eds, Fries R and Ruvinsky A), CABI Publishing, Oxon, UK, pp. 33–53.

Palombi, M and Damiano, C 2002, Comparison between RAPD and SSR molecular markers in detecting genetic variation in kiwifruit (*Actinidia deliciosa* A. Chev), *Plant. Cell. Rep.*, **20**, 1061–1066.

Pasqualone, A, Caponio, and Blanco, A 2001, Inter-simple sequence repeat DNA markers for identification of drupes from different *Olea europaea* L. cultivars, *Eur. Food Res. Technol.*, **213**, 240–243.

Pasqualone, A, Lotti, C and Blanco, A 1999, Identification of durum wheat cultivars and monovarietal semolinas by analysis of DNA microsatellites, *Eur. Food Res. Technol.*, **210**, 144–147.

Pasqualone, A, Montemurro, C, Caponio, F and Blanco, A 2004, Identification of virgin olive oil from different cultivars by analysis of DNA microsatellites, *J. Agric. Food Chem.*, **52**, 1068–1071.

Perry D J 2004, Identification of Canadian durum wheat varieties using a single PCR, **109**, 55–61.

Perry, M D, Davey, M R, Power, J B, Lowe, K C, Frances, H, Bligh, J, Roach, P C and Jones, C 1998, DNA isolation and AFLP™ genetic fingerprinting of *Theobroma cacao* (L.), *Plant Molec. Biol. Rep.*, **16**, 49–59.

Plieske, J and Struss, D 2001, Microsatellite markers for genome analysis in *Brassica*. I. development in *Brassica napus* and abundance in *Brassicaceae* species, *Theor. Appl. Genet.*, **102**, 689–694.

Prasad, M, Varshney, R K, Roy, J K, Balyan, H S and Gupta, P K 2000, The use of microsatellites for detecting DNA polymorphism, genotype identification and genetic diversity in wheat, *Theor. Appl. Genet.*, **100**, 584–592.

Ren, N and Timko, M P 2001, AFLP analysis of genetic polymorphism and evolutionary relationships among cultivated and wild Nicotiana species, *Genome*, **44**, 559–571.

Röder, M S, Wendehake, K, Korzun, V, Bredemeijer, G, Laborie, D, Bertrand, L, Isaac, P, Rendell, S, Jackson, J, Cooke, R J, Vosman, B and Ganal, M W 2002, Construction and analysis of a microsatellite-based database of European wheat varieties, *Theor. Appl. Genet.*, **106**, 67–73.

Rouzaud, F, Martin, J, Gallet, P F, Delourme, D, Goulemot-Leger, V, Amigues, Y, Menissier, F, Leveziel, H, Julien, R and Oulmouden, A 2000, A first genotyping assay of French cattle breeds based on a new allele of the extension gene encoding the melanocortin-1 receptor (Mc1r), *Genet. Sel. Evol.*, **32**, 511–520.

Ruiz, C, Paz Breto, M and Asíns, M J 2000, A quick methodology to identify sexual seedlings in citrus breeding programs using SSR markers, *Euphytica*, **112**, 89–94.

Sánchez-Pérez, R, Ruiz, D, Dicenta, F, Efea, H J and Martiníz-Gómez, P 2005, Application of simple sequence repeat (SSR) markers in apricot breeding: molecular characterization, protection, and genetic relationships, *Sci. Hort.*, **103**, 305–315.

Sasazaki, S, Itoh, K, Arimitsu, S, Imada, T, Takasuga, A, Nagaishi, H, Takano, S, Mannen, H and Tsuji, S 2004, Development of breed identification markers derived from AFLP in beef cattle, *Meat Sci.*, **67**, 275–280.

Saunders, J A, Hemeida, A A and Mischke, S 2000, USDA DNA fingerprinting programme for identification of *Theobroma cacao* accessions, in *INGENIC 2000 Proc. Int. Workshop New Technol. Cocoa Breeding*, pp. 108–114.

Savelkoul, P H M, Aarts H J M, Dijkshoorn, L, Duims, B, D E, Haas, J, Otsen, M, Schouls, L and Lenstra, J A 1999, Amplified fragment length polymorphism analysis: the state of an art, *J. Clin. Microbiol.*, **37**, 3089–3091.

Serrano, B, Gómez-Aparisi, J and Hormaza, J I 2002, Molecular fingerprinting of Prunus rootstocks using SSRs, *Hort. Sci. Biotech.*, **77**, 368–372.

Shim, S I and Jørgensen, R B, 2000, Genetic structure in cultivated and wild carrots (*Daucus carota* L.) revealed by AFLP analysis, *Theor. Appl. Genet.*, **101**, 227–233.

Sjakste, T G, Rashal, I and Roder, M S 2003, Inheritance of microsatellite alleles in pedigrees of Latvian barley varieties and related European ancestors, *Theor. Appl. Genet.*, **106**, 539–549.

Sobotka, R, Dolanska, L, Curn, V and Ovesna, J, 2004, Fluorescence-based AFLPs occur as the most suitable marker system for oilseed rape cultivar identification, *J. Appl. Genet.*, **45**, 161–173.

Terol, J, Mascarell, R, Fernandez-Pedrosa, V and Perez-Alonso, M 2002, Statistical validation of the identification of tuna species: bootstrap analysis of mitochondrial DNA sequences, *J. Agric. Food. Chem.*, **50**, 963–969.

This, P, Jung, A, Boccacci, P, Borrego, J, Botta, R, Costantini, L, Crespan, M, Dangl, G S, Eisenheld, C, Ferreira-Monteiro, F, Grando, S, Ibanez, J, Lacombe, T, Laucou, V, Magalhaes, R, Meredith, C P, Milani, N, Peterlunger, E, Regner, F, Zulini, L and Maul, E 2004, Development of a standard set of microsatellite reference alleles for identification of grape cultivars, *Theor. Appl. Genet.*, **109**, 1448–1458.

Tommasini, L, Batley, J, Arnold, G, Cooke, R, Donini, P, Lee, D, Law, J, Lowe, C, Moule, C, Trick, M and Edwards, K 2003, The development of multiplex simple sequence repeat (SSR) markers to complement distinctness, uniformity and stability testing of rape (*Brassica napus* L.) varieties, *Theor. Appl. Genet.*, **106**, 1091–1101.

Tyrka, M, Dziadczyk, P and Hortynski, V 2002, Simplified AFLP procedure as a tool for identification of strawberry cultivars and advanced breeding lines, *Euphytica*, **125**, 273–280.

Vázquez, J F, Pérez, T, Ureña, F, Gudín, E, Albornoz, J and Domínguez, A 2004, Practical application of DNA fingerprinting to trace beef, *J. Food. Protect.*, **67**, 972–979.

Verkaar, E L C, Boutaga, K, Nijman, I J and Lenstra, J A 2002, Differentiation of bovine species in beef by PCR-RFLP of mitochondrial and satellite DNA, *Meat. Sci.*, **60**, 365–369.

Verkaar, E L C, Vervaecke, H, Roden, C, Barwegen, M, Susilawati, T, Romero Mendoza, L, Nijman, I J and Lenstra, J A 2003, Paternally inherited markers in bovine hybrid populations, *Heredity*, **91**, 565–569.

Vidal, J R, Delavault, P, Coarer, M and Defontaine, A 2000, Design of grapevine (*Vitas vinifera* L.) cultivar-specific SCAR primers for PCR fingerprinting, *Theor. Appl. Genet.*, **101**, 1194–2001.

Viruel, M A and Hormaza, J I 2004, Development, characterization and variability analysis of microsatellites in lychee (*Litchi chinensis* Sonn., *Sapindaceae*), *Theor. Appl. Genet.*, **108**, 896–902.

Weder, J K P, 2002, Identification of food and feed legumes by RAPD-PCR, *Lebensm.-Wiss. u.-Technol.*, **35**, 504–511.

Werner, F A, Durstewitz, G, Habermann, F A, Thaller, G, Kramer, W, Kollers, S, Buitkamp, J, Georges, M, Brem, G, Mosner, J and Fries, R 2004, Detection and characterization of SNPs useful for identity control and parentage testing in major European dairy breeds, *Anim. Genet.*, **35**, 44–49.

Woolfe, M and Primrose, S 2004, Food forensics: using DNA technology to combat misdescription and fraud, *Tr. Biotechnol.*, **22**, 222–226.

Wünsch, A and Hormaza, J I 2002a, Cultivar identification and genetic fingerprinting of temperate fruit tree species using DNA markers, *Euphytica*, **125**, 59–67.

Wünsch, A and Hormaza, J I 2002b, Molecular characterisation of sweet cherry (*Prunus avium* L.) genotypes using peach [*Prunus persica* (L.) Batsch] SSR sequences, *Heredity*, **89**, 56–63.

Zeleny, R and Schimmel, H 2002, Sexing of beef – a survey of possible methods, *Meat. Sci.*, **60**, 69–75.

Zhebentyayeva, T N, Reighard, G L, Gorina, V M and Abbott, A G 2003, Simple sequence repeat (SSR) analysis for assessment of genetic variability in apricot germplasm, *Theor. Appl. Genet.*, **106**, 435–444.

Zheng, X, Song, S and Liu, H 2001, Identification of Chinese cabbage (*Brassica campestris* L. ssp. *Pekinensis*) cultivars with isozyme, RAPD and AFLP markers, *Acta Hort.*, **546**, 543–549.

Zietkiewicz, E, Rafalski, A and Labuda, D 1994, Genome fingerprinting by simple sequence repeat (SSR)-anchored polymerase chain reaction amplification, *Genomics*, **20**, 176–183.

9

Electronic identification, DNA profiling and traceability of farm animals

A. Poucet, C. Korn, U. Meloni, I. Solinas, G. Fiore and
M. Cuypers, European Union Joint Research Centre, Italy,
G. Caja and A. Sánchez, University of Barcelona, Spain, A.
Fonseca, P. Pinheiro and C. Roquete, University of Evora, Portugal

9.1 Introduction

The recent crises over Bovine Spongiform Encephalopathy (BSE) and Foot and Mouth Disease in Europe have demonstrated that a reliable traceability system is essential in order to ensure safe food. One of the most common weaknesses in the European Community (EC) Member States where these crises developed was the absence of rapid and effective ways of tracing animals and products that might have been contaminated. Without the ability to trace animals, it was not always possible to remove possibly contaminated animals and food from the market, thus consumer protection was not ensured. As a result, it was decided to introduce specific legislation concerning traceability that would oblige producers to provide the relevant authorities and consumers with clear information about the origin, processing and storage of meats.

Considerable research has been done to examine different methods of tracing meats, taking into account cost, efficiency, robustness and compatibility with modern food production systems. The electronic identification techniques described below are the preferred solution as a result of these studies. One of the main goals of the research was to help policymakers to improve legislation on food production, which is the only way to avoid further food crises and ensure consumer confidence. The recent EC Regulation 21/2004 provides Member States for the first time with the option to use electronic means to identify food producing animals classed as small ruminants. A similar regulation will be debated very soon for cattle, and serves to underline the value of work on electronic identification over recent years.

9.2   Electronic identification methods in tagging and traceability of cattle

9.2.1   Animal identification requirements

Livestock identification is based on a typical set of characteristics in an animal that allows a third party to differentiate quickly among a set of individuals of the same species, race and family[1]. The need to identify individual animals emerged with domestication, when it became necessary to identify ownership of animals in a flock or herd. Conventional systems of animal identification have evolved over time, and currently tattooing and ear-tagging are the most widely recognized. However, the disadvantages associated with these methods are no longer compatible with modern animal production.

According to Belda[1], a system of animal identification must be

- easy to implement
- easy to interpret
- difficult to counterfeit
- durable through all stages of the food chain and non-separable from the animal
- compatible with animal welfare

These requirements cannot entirely be met by a system based on conventional (non-electronic) ear-tagging and tattooing only.

The development of a method of identifying animals at a distance and in a short space of time must take into account the needs of the different stakeholders. These stakeholders and their requirements are described below.

- **Administration:** require a safe system of control for animal production and movement that cannot be counterfeited.
- **Food safety agencies:** require a means of rapidly identifying and controlling safety problems.
- **Breeders:** require better quality and flow of information about individual animals, flocks or herds, to improve identification, control of animal genealogies and aspects of animal production such as feeding regimes, animal weights and milk yields.
- **Breeders associations:** require better information for herd books and livestock registers, and the ability to certify products by geographical origin or method of production.
- **Slaughterhouses:** require automatic registration and identification of the origin of animals for slaughter, along with a means of tracking the animal beyond slaughter through permanent identification of the carcass, management of processing and quality control.
- **Consumers:** require access to a product's history to address safety concerns.

9.2.2   Development of electronic identification

Mandatory and optional information requirements in animal production are becoming increasingly exhaustive, not only to meet the need of producers to improve feeding, veterinary management and reproductive programmes, but also to meet the demands of authorities and consumers for safer and more reliable meat products. An effective system of animal identification must be safe, tamperproof, permanent and automatic. The Animal and Plant Health Inspection Service (APHIS) and Los Alamos National Laboratory (LANL) began the first studies on electronic identification of animals[2] in response to food crises in the second half of the 20th century. A range of options were investigated, including capsules with radioactive isotopes, magnetic bands, radio transmitters and optical devices. An electronic identification (EID) system based on radio frequency emerged as the most promising for further development[2].

Despite the clear advantages associated with the EID system, it also had to demonstrate commercial viability. Early applications for the industrial and livestock sector included necklaces and electronic keys used for automatic feeding and milking, the control of breeding programmes, and the management of pigs and cattle. Miniaturization and reduced production costs led to the manufacture of injectable transponders, but it was essential to demonstrate the reliability and functionality of the miniaturized system in practice. The International Committee for Animal Recording (ICAR) required a reading efficiency greater than 99%, as well as guarantees that the system was safe for both the animals and man, and that no toxic or dangerous components or residues remained in the edible parts of the animals after processing.

In 1993, the Commission of European Communities General Direction of Agriculture (DG VI) (FEOGA) developed a pilot project together with the Universidade Autónoma de Barcelona, Institute Zooprofilattico Sperimentale della Lombardia and dell'Emilia and Universidade de Évora. As part of fraud-prevention activities, the project looked at ways of solving the problems of identifying and registering livestock over the full range of production and developmental conditions in the EC. The main objective of the project was to study the wide-scale application of EID in the livestock species of interest. The project lasted one year, and transponders were implanted subcutaneously in 5000 sheep, 3000 bovines and 2000 goats. It was concluded that, in general, the electronic identification system was effective and sufficiently developed to be used under field conditions in commercial livestock operations.

The Identification Electronique des Animaux (IDEA) Project was conducted between 1998 and 2001, and investigated the potential for the Community-wide application of an EID system. The IDEA Project was carried out across six countries, Germany, Spain, France, Holland, Italy and Portugal, in a total of nine studies. The main objectives were to study the feasibility and evaluate the performance of an electronic identification system in ruminants (cattle, buffalo, sheep and goats), and identify the organizational structure needed to implement such a system for EC livestock. The IDEA Project demonstrated that livestock identification, registration and management across the Community can be substantially improved using an EID system.

9.2.3 The future

Animal production in Europe is undergoing major changes in response to measures introduced by the European Union itself and other international bodies such as the World Trade Organization (WTO or OMC), the Common Market of the Southern Cone (MERCOSUL), the North American Free Trade Agreement (NAFTA), and the Free Trade Area of the Americas (FTAA or ALCA). These measures are based on information flow, transparency and consistency of information across borders. In order to remain competitive, cattle production must introduce innovative methods of production and monitoring that will lead to improved quality, productivity and sustainability.

The recent BSE crisis placed unprecedented demands on the food security systems of the Community and individual Member States, and highlighted the deficiencies, particularly in traceability, that must now be addressed.

Traceability requires that livestock producers commit to providing identifying data for each animal that can be followed through the production chain. The IDEA Project has demonstrated that it is possible to implement an identification system based on EID that will result in substantial benefits to the meat-production sector. Introducing EID for livestock will lead to a major improvement in the quantitative and qualitative data available to manage animal production and to provide information to the consumer. It will also provide solutions to existing problems, both general in the case of cattle and more particular in the case of small ruminants. The EID system will not only provide a way of unalterably and uniquely identifying individual animals, but it will also enable the automated control and veterinary management of livestock. Automating data collection and linkages to a central database will save time and lead to increased confidence in the information provided to the consumer.

The IDEA Project has also shown that manufacturers of EID devices have the capacity to produce the equipment necessary to introduce EID across Europe, and that international standards for codification of transponders and communication protocols exist that will guarantee integration and compatibility.

The IDEA Project therefore demonstrated that the implementation of an EID system in goats, bovines, bubaline and sheep can be carried out in a timely manner, and will improve the efficiency of the identification, registration and management of livestock in the European Union.

9.3 Technical basis for animal identification by radiofrequency (RFID)

This section describes the technical characteristics and operating principles of electronic identification equipment for animals in the form of RFID using low frequency ISO compliant passive transponders[3]. Standardization to ensure overall compatibility, conformance and efficiency of electronic identification devices is discussed.

9.3.1   RF technology: operating principles of an electronic identification system

The transponder (TRANSmitter + resPONDER) is composed of a microchip, an air or ferrite coil and, in some cases, a capacitor. In order to read the internationally unique identification code[4] contained inside the transponder, the reader sends an electromagnetic activation field[3] at a frequency of 134.2 kHz. This charges the transponder, which then sends back its identification code to the reader when it discharges (see Fig. 9.1 and 9.2 below).

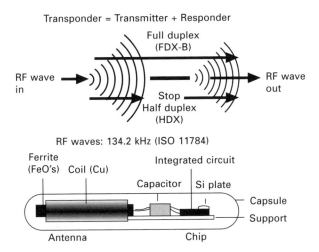

Fig. 9.1   RF technology – passive transponders.

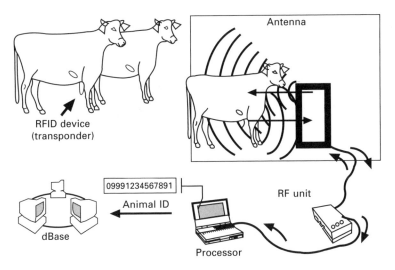

Fig. 9.2   RF technology – passive transponders.

There are two main transponder technologies, full duplex or FDX, and half duplex, or HDX.

Full duplex transponder
A full duplex (FDX-B) transponder (as defined in ISO 11785[3]) transmits its code during the activation period. The FDX transponder uses a modified DBP (differential bi-phase encoding) encoded sub-carrier which is amplitude-modulated. The transponder sends back its message using the frequency bands 129 to 133.2 kHz and 135.2 to 139.4 kHz.

Half duplex transponder
A half duplex (HDX)[3] transponder transmits its signal when the activation signal is interrupted. The HDX transponder responds between 1 ms and 2 ms after a 3 dB decay in the activation signal. The HDX transponder uses frequency shift keying (FSK)[3] modulation at (124.2 ± 2) kHz to transmit a binary 1 and at (134.2 ± 1.5) kHz to transmit a binary 0. The encoding signal is non-return to zero encoding (NRZ)[3], in which data bit 1 is a high signal and data bit 0 is a low signal.

9.3.2   Transponders

Three types of passive (containing no energy source) electronic identifiers for livestock are currently defined by the International Committee for Animal Recording (ICAR) and available on the market: the ruminal bolus, ear-tags and the injectable transponder.

The electronic ruminal bolus is composed of a cylindrical ceramic capsule which contains a passive transponder. This ceramic capsule is swallowed by the animal, using a mechanical applicator or so-called 'bolus gun' (such as that used to administer vitamins), and remains in the ruminant's fore-stomach, normally inside the reticulum (2nd stomach) as shown in Fig. 9.3.

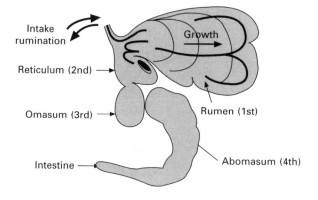

Fig. 9.3 Location of the bolus inside the animal.

Electronic ear tag transponders are plastic-covered transponders that can be fixed to the animal's ear using a locking mechanism in the same way as conventional plastic ear tags are applied, or attached in a non-reversible way to an ear tag.

The injectable transponder consists of a small transponder, encapsulated in a biocompatible and non-porous material (e.g. glass), that can be injected under the animal's skin using a mechanical applicator or injection gun equipped with a needle to perforate the skin.

In all three identifier types, the transponder has a unique identification number, the codification of which is discussed in section 9.3.5.

9.3.3 Transceivers (readers)

The passive electronic identifiers can be read using two types of readers or transceivers[3], portable and stationary.

Portable, hand-held readers run on batteries and are mostly used during the first identification of the animal. They are also used when animals are immobilized (e.g. restrained, tied in stalls, head lockers) and for small flocks. These readers have an antennae incorporated into the reader or can be connected to an external antenna such as a stick antenna when it is difficult to get close to the animal. Various models of portable reader are available on the market, such as the read-only reader, which only displays the identification code of the animal and does not have a memory. In this case, the identification code must be hand-recorded on paper. The identification code can be downloaded to a computer if the reader is equipped with an internal memory and a computer interface, and this type of low-cost reader is recommended in production units where identification codes are read occasionally. Programmable readers may include a keyboard, an internal memory and a link (cable or wireless) to a personal computer. This type of reader allows the use of previously recorded data related to the animal to be associated with a transponder identification code, such as date of birth, sex, breed, farm, and performance data. These data can then be transferred automatically to a personal computer and recorded in a database. Reference 5 contains a full list of devices certified according to the JRC test procedures.

Stationary readers (see Fig. 9.4) are mainly used for groups of animals, which are moved past the reader in a slaughterhouse or along a chute or raceway. A gate antenna, which can be shaped to accommodate operating conditions, can be installed on one side of the chute. The reading antenna is connected to a stationary radio-frequency reading unit, which is in turn connected to a data recording system such as a personal or hand-held computer. The animals pass through the chute in front of the antenna and the identification code is read and stored in the recording system, along with the time of reading.

Fig. 9.4   Example of a dynamic reading system with a raceway using a stationary reader.

9.3.4   Technical specifications for EID equipment

Three different levels of standardization are recommended to guarantee compliance with ISO standards and ensure compatibility: technical, operational and environmental. In Europe, a test laboratory (or a network of laboratories) will certify equipment that complies with the standards, so that Member States can ensure that they procure appropriate devices with guaranteed performance and compatibility.

Technical standards

Electronic identifiers must comply with the standards ISO 11784 and ISO 11785, which define the code structure and technical aspects regarding communication between transponder and reader. Electronic identifier readers have to comply with standard ISO 11785. This standard defines the technical concepts behind radio-frequency identification of animals and specifies that the reader must be able to read both identifier technologies, HDX and FDX-B.

Operational standards

Operational conditions include such things as species/size of animal and width of chute. These must be taken into account in order to guarantee full compatibility between identifiers and readers, and a minimum reading distance (under optimum orientation of the tags in respect to the antenna) is recommended.

Two different reading distances have been defined for the use of portable readers, depending on the type of electronic identifier used (injectable, ear-tag or bolus). The required reading distance for the ear-tag is shorter than that for injectable and bolus identifiers because ear-tags are always visible, and the operator can place the reader close to or in contact with the ear-tag if necessary.

Environmental standards

The quality and performance of electronic identifiers and readers must be checked under exposure to various stresses to allow for adverse operating conditions, such as temperature, humidity, mechanical vibration and electromagnetic disturbance (readers only). Examples of typical standardized test methods conducted during the IDEA Project[5] can be obtained from its web site (http://certificazioni.jrc.it).

9.3.5   Code structure of the transponders

The basic function of the codification system is the capability to trace back, through the identifier code, the basic data (e.g. information on date and place of origin) and history (movements throughout the Community) of an animal from birth to slaughter. The link between the identification code and the basic data and the history of the animal is made through the local and/or national animal registration system. When an animal is moved to another holding within the same country or to another country, the unique identification code of the animal, together with the movement document, should identify the database where information relevant to the animal is available.

To create unique electronic identifiers, the code structure of the transponder has to fulfil the specific requirements of version 2004 of ISO 11784, as summarized below.

- The first four positions identify the Member State where the animal is first identified. For this purpose, four digits, starting with a numeric zero, and then followed by ISO 3166 (3 numeric) country codes.
- The 12 characters after the country code, the animal identification number, shall be numeric and the number of combinations shall not exceed 274.877.906.944.
- The basic code structure has to be compatible with ISO standard 11784 (version 2004).

In practice, the transponder code is structured with different parts, as illustrated in Table 9.1 (ISO 11784 version 2004).

Bits 2 to 4

These provide information on the retagging of the animal. In Europe, Regulation 21/2004 concerning sheep and goats in article 4 point 6 states that:

> No means of identification may be removed or replaced without the permission of the Competent Authority. Where a means of identification has become illegible or has been lost, a replacement bearing the same code shall be applied in accordance with this article. In addition to the

Table 9.1   Structure of the identification code for electronic animal identification (ISO 11784 version 2004)

Bit (s) n°	Number of digits	Description
1	1	This bit indicates whether the transponder is used for animal identification or not. In all animal applications this bit shall be 1
2–4	1	Mandatory retagging counter
5–9	2	Mandatory species code (0 to 31)
10–15	5	Empty – all zeros (reserved zone for future applications)
16	1	Bit indicating the presence or not of a data block. 0 = (no data block)
17–26	4	Mandatory ISO 3166 numeric 3-digit country code prefix zero
27–64	12	0 – 274.877.906.943 National identification code. (unique number in each country) - 274.877.906.944 combinations

code and distinct from it the replacement may bear a mark with the version number of the replacement.

This last point is implemented in the following manner:

- At the first application of an electronic identifier to an animal, the number (retagging code) has to be set to 0.
- When the same animal has to be retagged (loss or failure of the original electronic tag), this number should be set to 1 for the first retagging, 2 for the second retagging, etc. It is possible to retag the same animal up to seven times, if necessary, by changing the retagging counter for each new electronic tag in the data bits 2 to 4. If the animal has to be retagged more than 7 times, authorization must be requested from the competent authority in order to change the National Identification Code in bits 27 to 64.

Regulation No. 21/2004 in article 4 point 6 also states that an animal can be re-tagged with a new identification number, in this case, the re-tagging code is reset to 0.

Bits 5 to 9
In order to guarantee a unique identity for any animal, independent of the species, the possibility of including a species code in the transponder code structure has been agreed with ISO technical working groups. The bits reserved for the species can contain a maximum of 31 different codifications.

Existing codifications could be used for animal species under the existing Combined Nomenclature (CN) rules. These were established to implement Council Regulation (EEC) No 2658/87 of 23 July 1987, which established the basic principles of codification and a first list of custom codes aimed at the unique identification of goods for custom purposes. According to these rules, live animals can be classified as shown in Table 9.2.

Table 9.2 CN codification of species and transponder codification (bits 5 to 9)

CN Code	Transponder species code	Animal species
	00	No species code defined
0101	01	Live horses, asses, mules and hinnies
0102	02	Live bovine animals
0103	03	Live swine
0104	04	Live sheep and goats
0105	05	Live poultry, that is to say, fowls of the species Gallus domesticus, ducks, geese, turkeys and guinea fowls
0106	06	Other live animals

Owing to current constraints on the new ISO 11784 transponder codification (only 2 digits are available for the identification of animal species), the first 2 digits of the CN code can be removed and the species codified using the abbreviated codification in bits 5 to 9, also shown in Table 9.1.

It is clear that a harmonized approach is needed for the codification of transponders in the different countries of the EU. Unique identifiers at EU level are needed to ensure that animal transfers within and between countries are fully transparent and compliant with traceability requirements.

9.4 EID equipment for animal identification on farms and in slaughterhouses

9.4.1 Electronic ruminal boluses

As described in section 9.3.2 above, electronic ruminal or reticular boluses are composed of a ceramic capsule that contains a passive transponder. An applicator is used to place the bolus at the back of the animal's tongue, which causes involuntary swallowing. The bolus then remains in the rumen or reticulum for the lifespan of the animal.

The process of identifying an animal with a bolus can be divided into five stages:

- test reading of the bolus, before application;
- application of the bolus;
- using the transceiver to read the bolus and confirm correct application (the most likely position for the rumen is on the left of the animal, so the reading should be taken on this side);
- inputting the individual animal identification data into the reader; and
- transferring the identification data to a computer and subsequently onto the appropriate database.

Application may be carried out by one operator working alone, for smaller animals, or with an assistant for larger animals. The following equipment is necessary:

- boluses with integrated transponders, with reading (R/O) and with identification number programmed by the manufacturer (ISO 11784);
- bolus applicators;
- portable programmable reader for the introduction of animal data;
- portable computer with the appropriate software;
- printer.

To apply a ruminal bolus, the animal must be restrained, typically in a restricted space, to reduce mobility to a minimum (Fig. 9.5 and 9.6). For bovines a chute is appropriate for sheep or goats, a chute or pen can be used. Care must be taken to safeguard the well-being of the animal and keep it calm, as well as ensuring that the operator can apply the bolus quickly and efficiently. The restraint must allow the animal's head to remain in its natural position, while permitting extension of the neck in order to help swallowing. It is recommended that the nostrils are not raised above the line of the eyes when introducing the applicator and that sideways movement of the head is restricted. This type of application must not be carried out on animals that are lying down, and should not be carried out on unhealthy animals. Animals that must be stood up or moved to a suitable location for application should be allowed to rest before application is carried out.

Once the animal is restrained, the applicator is introduced into the mouth at the side, between the lips and towards the back of the lower jaw. If the animal does not swallow the bolus involuntarily, this may lead to the bolus

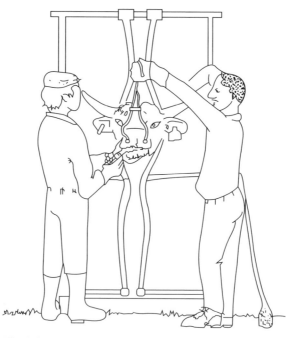

Fig. 9.5   How to apply the bolus (viewed from the front).

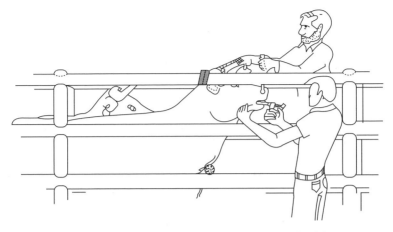

Fig. 9.6   How to apply the bolus (viewed from the side).

entering the animal's windpipe and interfering with respiration. If this occurs and the animal appears to be choking, the neck should be allowed to extend and a firm pat should be given in the region of the nape of the neck.

The technique for application depends on the animal species. In the case of small ruminants, the operator will work inside the area where the animals are restrained. If the operator has an assistant, their job will be to immobilize the animal's head, and the assistant should approach the animal from the side (Fig. 9.7).

Fig. 9.7   A second example of how to apply the bolus.

One of the best methods of restraint for an operator working alone consists of placing the neck of the animal between the operator's legs. By keeping his or her legs in front of the animal's shoulders, the operator can restrain the animal and apply the bolus quickly and efficiently. All materials used in the application of ruminal boluses must be washed and disinfected before being used for other activities.

To apply ruminal boluses in bovines, the operator needs an assistant to restrain the animal and ensure the correct head position. The heads of animals with horns can be restrained by lassoing a rope to the horns, which is then coiled over the superior transversal bar of the chute. In the case of animals without horns, a trunk or feeder equipped with a containment system (Dutch or American type) can be used to immobilize the head. Once the head is restrained, the operator should encourage the animal to open its mouth by inserting one or two fingers between the lips to one side of the head. At the same time, the operator should introduce the applicator between the teeth and apply the bolus to the back of the tongue.

9.4.2   Reading devices for on-farm reading

Electronic identification data can be read with a portable or a fixed reader. Fixed readers, used to identify moving animals, comprise a reader, an antenna and a computer. The equipment is set up in a corridor and the animals pass through one at a time. Having the appropriate software on a portable computer means that animals can be counted or selected out of a flock or herd and comparisons with earlier readings can be made to identify missing animals where necessary.

Portable readers can be used in a variety of situations, not just when animals are stationary. Portable readers can be used to complement fixed readers where there is doubt, or where a small group of animals are kept separately (for instance expectant mothers or newborns). It can also be used where physical or management conditions do not permit the use of a fixed reader, such as when there is no corridor to facilitate single-file movement of the animals. Again, the appropriate software is necessary to make use of the readings.

To take readings from moving animals, the following are needed:

- a chute;
- fixed readers (reader + antenna);
- portable computer with appropriate software;
- portable printer;
- portable reader for comparison and verification of readings.

The fixed reading equipment must be set up correctly to ensure accurate readings. Placement of the equipment in the corridor should allow for an appropriate rate of movement of the animals, and ensure that interference from other equipment (e.g. mobile telephones, motors) is minimized. It is also important to have space in the corridor to allow for the appropriate action when an animal does not give a reading. The distance to the reader, the level of interference and the circulation of the animals should all be assessed before the readings are carried out.

9.4.3   Reading devices for slaughterhouses

It is possible to automate readings in slaughterhouses to follow the progress of an animal from when it enters the slaughterhouse to slaughter and evisceration. A series of readers linked to a computer can be set up to collect the relevant data at the appropriate time, and automatically update the central database so that slaughtered animals are registered as dead.

As with any other dynamic readings, setting up the readers to ensure optimum conditions for data gathering is important. In a slaughterhouse, there are additional conditions, such as temperature, humidity and electromagnetic interference, which must be taken into account.

The recovery of the bolus after evisceration of the animal in a slaughterhouse is relatively easy, and should be carried out in the 'dirty zone' of the line (guts zone), where there is no risk of contaminating the animal carcass. The gastric compartments are kept together, along with the intestines and the terminal portion of oesophagus. By identifying the oesophagus, its point of entry into the stomach can be located, and hence the position of the reticulum. The identity of the reticulum can be confirmed by cutting it open to reveal the characteristic honeycomb folds in its epithelium. It is usually possible to locate the bolus by touch, but if this is not the case, a reader can be used to find its position. Bolus recovery from slaughterhouses was determined to be easy and efficient during the IDEA Project, and did not require specialized personnel.

9.5   Data management

The introduction of electronic identification to track livestock will eliminate or significantly reduce the need for manual data entry. Using the equipment and methods described above will enable an animal to be followed from its original tagging to eventual slaughter, including movements within and between countries. An additional benefit will be the opportunity for quantitative analysis of the data, but this will only be possible if all data recorded are consistent.

In order to ensure that the data recorded are consistent, the database must be carefully structured and modelled, taking into account the type of data being recorded. A relational database for animal tracking through electronic identification will be composed of the following four sub-types of data:

* administrative data;
* animal identification data;

- animal movement data;
- point of recovery data.

Syntactic and semantic criteria will allow automatic validation of data that are loaded onto the database, and ensure that all data loaded are coherent. Figure 9.8 shows the steps from receiving data to recording on the main database.

On receipt, a message containing data to be recorded in the database is analysed to define its origin and consistency. Once authenticated, the data are processed and subjected to a syntactic check, and then stored in the buffer. If the message does not pass the syntactic checks, it is returned to the sender with the reason for non-acceptance.

The data are then loaded from the buffer to the temporary database. During this process, semantic and consistency checks are performed. If validated, the data are inserted into the database.

There are three options for data under this system:

- wrong data: returned to the subcontractor to be corrected and re-sent to the database;
- correct data: validated and recorded on the database;
- inconsistent data: correct data that are not recorded on the database because they are not coherent with existing data. Inconsistent data are placed on standby status, and when the complementary data are sent, all data are checked again and finally recorded.

Rejected data messages are returned to the originator, together with a complete report on all treatments performed during this phase.

The temporary database is a working database, and data are only transferred to the main database when they are consolidated. The main database is the reference database, and all statistical analyses are performed using data from the main database.

The IDEA Project has identified a number of technical requirements for data registration, transmission and management, which must be addressed as a matter of high priority to further develop the identification and registration system. Recommendations for addressing these are:

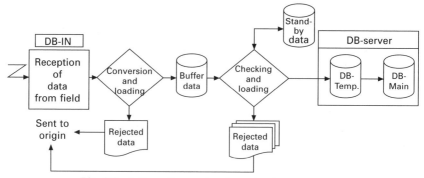

Fig. 9.8   Information system for animal management.

1. The syntactic rules for setting electronic instruments used to read animals (format/data/time) must be defined so that there are no discrepancies, otherwise such discrepancies could produce major errors in the registration of data on the databases.
2. The reliability (quality) of the data generated at the farm is the fundamental basis for greater efficiencies in livestock data management. Therefore, the introduction of electronic identification must be accompanied by a large effort to improve the management of farm registers by introducing modern information technology, e.g. programmable readers, electronic passport data transmission methods. The modernization of farm registers can be achieved through the efforts of individual farmers and/or through professional organizations, which are often already in charge of control and data management of animals in associated holdings.
3. Transport is one of the key elements in the overall animal identification, registration and management system, because it is the basis for locating and tracing animals in different holdings and slaughterhouses. Improvement of the management and data registration system of animal transport is necessary.
4. A databank should be created at EU level, containing the common glossary, data dictionary and its codification.

9.6   Future trends

The implementation of electronic identification will be an important step, as it will introduce improvements in the efficiency of the identification system, registration and management of European Union cattle stocks. Creating an EU-wide system will be a major challenge, particularly among newer Member States.

In the first phase, the concepts and definitions invoked in the new regulation (Reg. No. 21/2004), so that consistent procedures can be introduced at the European and national level, will be discussed and clarified.

A competent authority, defined for each country, will be involved across the range of activities, from sanitary administration to producer groups, slaughterhouses and consumers. All sectors must be involved, using appropriate communication channels (such as periodicals, fairs, congresses, television programmes, pilot projects) to inform and motivate.

It will be necessary to:

- guarantee the coherent management of data by developing a glossary of common terms, a dictionary of data and protocols of communications standards;
- take into account the diversity and status of animal production systems, along with the different types of equipment available, in drawing up identification criteria;

- obtain cooperation among Member States, and between Member States and the European Commission, for the development, periodic revision and improvement of the animal identification system.

9.7   References

1. Belda, A 1981, Identificacion Animal. Ministerio de Agricultura, Publicaciones de extension agrária, 3° ed., Madrid.
2. Spahr, S A 1992, Methods and systems for cow identification. In: *Large dairy herd management*. Van Horn, H H and Wilcox, C J (Eds). Management Services, American Dairy Science Association. Champaign, Illinois.
3. ISO 11785: 1996, Radio-frequency identification of animals – Technical concept.
4. International Committee for Animal Production (ICAR), International Agreement of Recording Practices, 2005.
5. http://certificazioni.jrc.it

10

Storing and transmitting traceability data across the food supply chain

R. Vernède and I. Wienk, Wageningen University and Research Centre, The Netherlands

10.1 Introduction

Optimisation of production, storage and distribution processes within chains is one of the continuous interests of commercial companies. Optimisation can refer to a wide range of aspects such as optimal logistics, high quality, low stocks and, for food producing companies, highest standards of food safety. In order for an individual company or an interconnected chain to improve performance, accurate and up to date information on the status of products is essential. Modern data carrier technologies in combination with modern ICT facilities are already able today to provide such information.

Depending on the goals companies and chains want to realise, appropriate data technology will have to be selected (see also FoodPrint method as presented in this book in Chapter 2). Companies and chains normally start with identification of products. Later, they progress to more advanced applications and start to utilise tracking and tracing also for other purposes. Thus, the conditions under which the products are stored and transported, for example, are monitored. In the future, they may wish to measure the quality of products based on the intrinsic characteristics of the product. Figure 10.1 visualises the different levels of data carrier technologies.

The main components of a tracking and tracing system are presented in Fig. 10.2. The major tracking and tracing modules are (1) data carrier technology, (2) data collection and (3) data processing. Supporting infrastructure refers to topics like the harmonisation of bar-coding, network- and web-interfaces and arrangements on data- and product-ownership, transparency, and liability between chain partners. Each of the technology components will be elaborated in detail. In the following sections, the three types of data

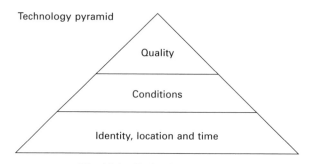

Fig. 10.1   Technology pyramid.

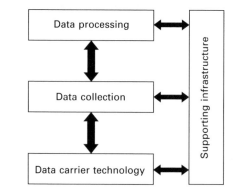

Fig. 10.2   Technology components of a tracking and tracing system.

carrier technologies for identification, condition measurement and quality measurement will be discussed followed by a short introduction to data collection and processing.

10.2   Product identification data carrier technology

In order to track and trace individual products or batches of products, these need to be made identifiable. Technologies for identification can be grouped according to the method by which the encoded data is stored (AIM, 2002). The following three main groups can be distinguished: (1) optical storage, (2) magnetic storage and (3) electronic storage. Now, a fourth category can be added: biological storage. This last category, in which the identity is measured by some aspects of the make-up of the specific product, is also referred to as primary identification. See Fig. 10.3 for a more detailed clustering of data carrier technologies (AIM, 2002).

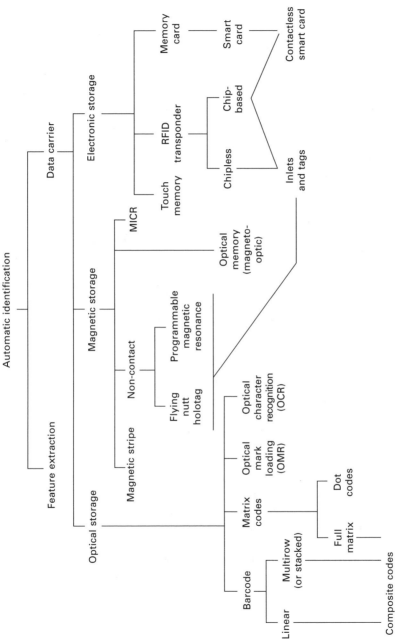

Fig. 10.3 Data carrier technologies (Source: AIM, 2002).

10.2.1 Optical storage

The simplest way of identification is by labelling the product unit with an alphanumeric label. This label serves as key in the batch administration. Different types of barcode labelling are also used: linear and multi-row barcodes and matrix codes. For barcodes, most commonly the EAN-UCC standard is employed for standardisation thus facilitating electronic data interchange (EDI). The GTIN (Global Trade Item Number) (EAN 13) forms the basis for worldwide identification through bar coding. The more advanced EAN 128 code (symbology and application identifier standard) makes it possible to include additional information such as expiry date, batch or serial number. Moreover, with a SSCC (Serial Shipping Container Code), transport units (pallets, containers) can be identified worldwide using the EAN 128 code. Optical coding can also take place in the product itself (for example 'carved' tagging of meat).

10.2.2 Magnetic storage

The best known examples of this group are the commonly used bank cards. However, most cards today are not only magnetic but also contain an electronic chip (smart chip). Magnetic strips are also commonly used for boarding cards at airports for the purpose of identifying passengers quickly and other security access purposes. The information on magnetic card can easily be damaged or erased under the influence of strong magnetic fields.

10.2.3 Electronic storage

This group includes smart cards, touch memory and RFID. An RFID tag, also known as a transponder, is a small microchip with an antenna. In reaction to a radio signal, the chip performs a simple process (e.g. send the ID-code). In order to read the tags, reader-antenna units are required. RFID tags are divided into passive (without a battery) and active tags (with a small battery). A critical aspect for wide-scale introduction of RFID is the need for standardisation of the used radio frequency. Currently, different frequency areas are in use: low (<1 MHz), medium (1–500 MHz) and high (>500 MHz). The higher the frequency, the longer the reading distance is, varying from only a couple of centimetres up to several hundreds of metres for high-frequency active tags. The approved frequencies and sending capacities differ between Europe and the rest of the world. Standardisation is currently in progress under the umbrella of EPC Global who are setting up a world-wide Electronic Product Code (EPC).

10.2.4 Biological storage

Bio-tagging, the exploitation of 'biological' characteristics for identification purposes, is a next step in the development of tagging technology. Biotagging comes in various forms: *active* and *passive*, *natural* and *synthetic*, and tagging based on *physical*, *biochemical* and *genetic* information carriers. Active biotagging occurs when the tag is applied by humans, an example being the use of antibodies for (synthetic) peptides as biological 'barcode' in animals (Urlings, 2002). Passive biotagging, in contrast, applies biological mechanisms that are inherently present in biological material, for instance the use of skin patterns (Friesian pedigree cattle) and DNA fingerprints when identifying livestock (Agriholland, 2003). The contrast between natural and synthetic often, but not always, coincides with passive versus active. The information carrier is physical when the product itself is used as carrier of a code. Biochemical tagging deploys biochemicals as code carriers, as is the case in the example of tagging livestock with immunological codes, while genetic tagging deploys DNA-related material (including RNA and others) to carry the code, as is the case in the DNA fingerprinting example.

Combining several approaches may increase safety, and give the best of all worlds, an example being the combination of a barcode and RFID tag. Some actors in logistic chains, e.g. couriers, read the barcode, while others access the identical information through the RFID tag, allowing massive and fully automated processing in warehouses.

10.3   Condition measuring data carrier technology

Apart from identification, various, mainly environmental, conditions of products can be measured. The most important with regard to agricultural products are the following:

- *Temperature* of product or air around product
- *Relative humidity* of product or air around product
- *Gas concentration* around product (such as aromas for fruits or specific chemical components indicating decay of meat and fish)
- *Vibration* or *shock*, for example in the case of transport of sensitive flowers or fruits
- *Opening* or *closing* of the door of container, truck or storage room
- *Movements*, for example in the case of unauthorised movement of products
- *Direction* of a product, for example when products needs to be kept upright
- *Location* of product during transport.

The various available devices can be divided into two main types: chemical (including enzymatic) units and electronic units.

10.3.1   Chemical units
- *Time–temperature integrators/indicators (TTI):* TTIs are stickers which allow monitoring of temperature abuse, e.g. when a certain temperature limit is reached its colour changes. TTI are based on chemical, enzymatic

or physical reactions. The higher the temperature, the quicker these processes occur. The simplest version is the Temperature Indicator (TI), which changes colour above a predefined limit. More sophisticated are the TTIs which change colour when the temperature limit is exceeded for a certain period. Two types can be distinguished: (a) partial history time–temperature indicators: these indicators become active above a certain temperature limit and integrate the time a product has been above the threshold and (b) full history time–temperature indicators: these indicators start integration of the temperature starting from the point they are activated. TTIs have to be designed for the application by the producer and have to be activated by, for example, pressing or UV light.

- *Humidity indicators:* Similar to the TTIs, stickers are available which record the maximum relative humidity (RH) during storage and transport of moisture-sensitive products. Different colours indicate specific humidity ranges. The stickers are based on blue crystals which dissolve when exposed to the rated RH for a long period.

- *Leak indicators:* A leak indicator attached to a modified atmosphere package is able to ensure the package integrity throughout the distribution chain. A leak indicator may be an O_2 or a CO_2 indicator. A redox dye present in an O_2 indicator changes colour after oxidation with O_2. A typical CO_2 indicator contains several indicator strips. Each strip has been designed to change colour only when the CO_2 concentration is below a certain limit. The concentration of CO_2 is indicated by a colour change in one or more of the strips. (Smolander, *et al.*, 1997).

- *Ripening indicators:* A ripening sensor changes colour by reacting to the aroma released by fruit as it ripens. For pears, a sensor is already commercially available (www.ripesense.co.nz).

- *Biosensors:* There is a huge market for rapid detection biosensors that indicate the presence of pathogens in food packages (Alocilja and Radke, 2003). Research breakthroughs in this field are protected in patent applications (Jorma *et al.*, 2003).

- *Tilt indicators:* The tilt indicator is a sticker which changes colour if the product is not kept upright.

- *Shock indicators:* Stickers available in different sensitivity categories, which change colour when a shock is exposed to the product above the activation level.

10.3.2   Electronic units

- *Fixed sensors:* Fixed sensors in storage rooms, truck and aircrafts connected by cable or wireless transmission by means of GPRS to a network give information on the condition not of a product but of a certain location. If one knows, however, which products have been at that location for which period one obtains information about the product conditions as well.

- *Data loggers:* Data loggers are small devices with an independent power supply measuring conditions such as temperature, humidity, light, sound, certain gases. Some possess a small display indicating measured conditions and others work with sound or flashing light when a threshold is exceeded. A very wide range of data loggers is available with regard to size, price, capacities and possibilities for measuring conditions. Data loggers are normally programmed by the users themselves. Collected information can be retrieved from the data loggers by cable, IR, light or RF. Some data loggers are intended for single use, others for multiple usage.

- *i-Buttons:* Similar to data loggers are the contact memory buttons or i-buttons, which are small, very robust, devices in stainless steel and are capable of measuring temperature. Programming and reading data takes place by a short galvanic contact with the button.
- *RFID*$^+$*(Smart tags):* RFID$^+$ are active RFID tags, which have an embedded small sensor measuring for example temperature, light intensity or movements. Some RFID tags are able to build up a small network and to communicate to each other though a hub to the receiving antenna. Based on the strength of the signal, the relative location of the tags can be determined.

10.4   Quality measuring carrier technology

Direct and preferable online quality measuring of food products is hardly possible at the moment. Even though significant progress has been made with regard to genomics. Genomics is the study of the bio-molecules of living objects and gives direct insight into their control system. Results of cropping conditions as well as for example transport or storage conditions or first signs of quality detoriation can be monitored based on the gen-activities and expression. Genomics can be split up into different sub-disciplines: transcriptomics, proteomics and metabolomics. This division is made based on the type of substances which are investigated: messenger molecules (mRNA), proteins or metabolics. Each of these contain relevant and detailed information on the actual status of the product. At the moment, no ready-to-use devices based on genomics are available. Extensive studies are required and one of the major points for attention is the interpretation of the biological variance between different product of the same species.

10.5   Data collection

In order to make products identifiable and traceable the products need to be equipped with a tag, which can be tracked through the logistic chain. Relevant data needs to be registered and administered into a database. The registration method is strongly connected with the applied identification technology.

Moreover, requirements in organisations may further determine which of the following registration tools are used.

1 Alphanumeric labels many be registered by humans (keyboard, voice response technology) and automatically using an OCR reader. Barcodes can be read manually, using a hand-held device, or with automatic reading devices. Optical contaminations caused by dirt or physical damage to the tag hinder reading. Physical tags, such as the Bi-coder and Dot-code on carcasses, can be read automatically, deploying a vision system.
2 Magnetic tags can be read with relatively simple reading devices. However, dirt affects the reading capabilities.
3 RFID tags can only be registered by using a reader–antenna combination. These vary from small hand-held devices to fixed stations. Setting up a running and stable RFID system requires the effort of specialists.
4 Reading bio-tags is more complex and strongly depends on the applied label technology, while DNA code requires the intervention of a laboratory. Some of the biochemical codes are, or are expected to become, readable with so-called dipsticks.

10.6   Data processing

Once items are correctly identified, registered and administered into a database, information needs to be processed and analysed. With regard to data processing of tracking and tracing information in a supply chain, the following scenarios are distinguished (see also Fig. 10.4).

10.6.1   Distributed data storage and processing
Chain actors maintain their own data and exchange data on request by means of product batch identity.

10.6.2   Centralised data storage and processing
Information is stored and processed in a central database. Chain actors have access to the subset of information relevant to their business. Security of the central database is crucial. The central infrastructure can be maintained by a dominant chain actor (the chain director) or by an independent facilitator (trusted third party). Web-based applications are gaining popularity, especially when many small and medium-sized actors are involved (Wilson and Clarke, 1998). Larger chain actors may deploy interfaces such as XML (Extensible Markup Language) between their own ERP system (Enterprise Resource Planning), WMS (Warehouse Management System) or LIMS (Laboratory Information Management System) and a central database. In an emergency,

Option (a): Distributed data storage and processing

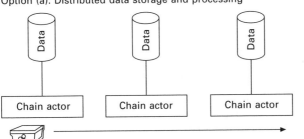

Option (b): Centralised data storage and processing

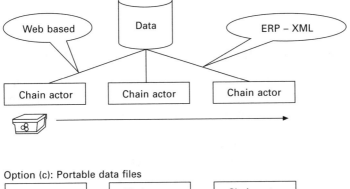

Option (c): Portable data files

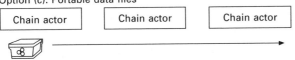

Fig. 10.4 Possibilities for data processing (Source: Vernède *et al.*, 2004).

the key-holder is authorised, in conformity with laid-down procedures, to access the database and extract relevant information.

In the food production chain, a combination of options (a) and (b) is common especially for products 'entering' another chain such as processed meat which is used in pizzas or frozen ready-to-eat meals. In these cases, the key-holder is authorised in specific cases to access a decentralised database of an individual company.

10.6.3 Portable data files
All relevant information travels physically through the supply chain, on paper or electronically on, e.g., RFID tags. Step by step information is added. In this scenario, intelligence is placed at the lowest level. This approach is, in practice, often combined with central systems as mentioned before.

The collected data then needs to be interpreted. For example, for a recorded temperature profile during the international airfreight transport of flowers, conclusions will have to be drawn with regard to the remaining shelf-life of the flowers. How does a short peak of high temperature effect the shelf-life compared with a longer but lower peak? Using a quality-gradient model for the specific product, predictions can be made. This model has to be based on experimental data obtained under standardised conditions. Unfortunately not many prediction models are available today.

The intelligence with regard to the quality-gradient model can be centralised, but it is also possible to have a data-logger or RFID tag incorporated within the model. By means of a small display or sound information, thresholds can be directly communicated to the user. Some TTIs already have preset thresholds for specific purposes. The colour changes when a critical temperature value is passed.

10.7   Practical applications of data carrier technology

From the previous section, it is clear that a wide range of data carrier technology is available to support quality-oriented tracking and tracing. Depending on the situation and the demands of the individual company or an interconnected chain more or less detailed information can be gathered and communicated. Using various scenarios, possible applications are discussed and evaluated on their specific strengths and weaknesses. Technical information is converted into practical cases, which will give an impression on how quality-oriented tracking and tracing is already used today or will be realised in the (near) future. The main focus of the scenarios is on the condition and quality-measuring technology. Identification technology and type of data processing have been chosen to match the sensor technology of the scenario (see Table 10.1).

Table 10.1   Overview of technology scenarios

		Technology scenarios				
		A	B	C	D	E
Identification	Optical	X	X			
	Magnetic					
	Electronic				X	
	Biological					X
Conditions	Colour labels	X				
	Fixed sensor		X			
	Data logger			X		
	i-Button					
	Smart tags				X	
Quality	Genomics					X
Data processing	Distributed		X			
	Centralised			X		X
	Portable	X			X	

Control of storage and transport conditions is of major importance to ensure the quality and safety of food products in general and perishable products in particular. Temperature is generally known to be the most important condition. Temperature strongly affects processes leading to quality decay and these effects have been studied in detail for many food products. Most of the technology scenarios of this section will therefore be focused on detection and registration of temperature.

10.7.1 Scenario A: Colour labels as alternative for best-before-date

In supermarkets, the main tasks for personnel are to assist clients at checkout and keep the shelves filled with products. For product information, customers have to cope with the text printed on the packaging of the products. Quality of many food products is communicated by means of the best-before-date. As long as the best-before-date is not passed, the producer claims that the quality of the product is good and consumption is safe. Instead of, or additional to, a best-before-date, a colour label can be placed on packed products.

TTI (Time–Temperature Indicator) labels placed on meat products that are stored at the optimum temperature will change colour at the moment the best-before-date is reached. When there has been a temperature abuse, the colour change is earlier in time as the shelf life of the products is reduced. In this case, the information on the product is not consistent which will raise a lot of questions. For this reason, the technology has not yet been widely embraced by retail, although it has been available for years. However, if the temperature conditions in the distribution chain are well controlled, this technology offers supermarkets an affordable method to inform customers about product quality and strengthen consumer trust.

Another type of colour label with great potential is the ripening indicator. Placed on a closed pack containing pears, the ripening indicator changes colour when the concentration of ripening aromas has surpassed a certain level and the pears are ready for consumption. This will prevent buyers from damaging fruits when they touch and squeeze them to judge the ripeness.

The main disadvantages of colour labels are that each type of product needs a dedicated label and extra handling or investments are required to place and activate the indicators on the packaging. Customers might have problems interpreting the information indicated by the colours, as they have to compare the actual colour of the label with the printed colour index. Colour labels are not suitable for use in the distribution chain since sharing of visually read information is complicated. Also, with TTIs it is not possible to determine when the temperature abuse has taken place, which is the main reason for applying sensors in distribution chains as will be outlined in section 10.7.2.

10.7.2   Scenario B: Sensors monitor closed cool chain

It is common knowledge that raw meat is stored best at reduced temperatures. Every kitchen is equipped with a fridge for that reason. Typically, the best-before-date of these products is based on studies of bacteria growth on meat at 4 °C, after which a certain safety margin is added. This implies that the use-by-date is a reliable quality parameter only if the meat is kept at 4 °C during the whole storage and transport period. Cold stores and trucks for conditioned transport are equipped with cool units. The air temperature in the locations can be measured using fixed sensors preferably placed at critical spots for instance near the doors. The sensor reading is used to automatically regulate the cooling equipment or to alarm personnel in case a threshold value has passed. Data are stored for a certain period to serve as evidence in case the temperature control of the location is questioned. These actions are part of quality management systems such as HACCP and BRC.

Proactive communication of temperature conditions to other partners in the distribution chain is not obvious and generally not practised. Yet, a partner sharing this type of information clearly takes his or her responsibility for product quality and can avoid claims.

Technical innovations are not necessary to link temperature data of the location to the products in that location. For plain logistic reasons, all items entering or leaving a warehouse or transport unit are registered. In a computer or a network surrounding the item, identity can easily be extended with information about its residence time at a certain location and the temperature of that location.

A gap in this quality system is the lack of temperature data at transfer points. In particular, platforms at airports are notorious spots where goods can occasionally be kept at higher temperatures (in the sun) and for longer periods than initially intended and agreed upon. To fill these gaps of information data loggers or portable data files have to be used as will be outlined in the next section.

10.7.3   Scenario C: Chain conditions validated with data loggers

For shipment of perishable food products conveyors, shippers and airlines are carefully selected. Most companies have updated and certified quality management systems. Goods are delivered together with certificates indicating that transport has occurred at agreed temperatures. But often certificates refer to intended conditions not actual measured data (see Fig. 10.5). Will these shippers inform their customers what has happened to the temperature of the products when the truck driver turned down the cooling equipment because the noise kept him or her from sleeping? Or when a flight is delayed or when customs formalities took much longer than expected? Occasional validation of transport conditions such as temperature and relative humidity gives valuable and complete information of the actual transport conditions including transfer points. Data loggers are the preferred sensor types for this

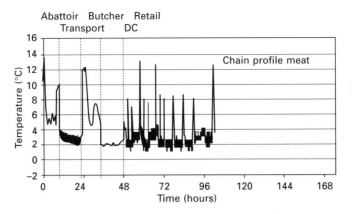

Fig. 10.5 Illustrative data set of temperature in a meat chain from abattoir to retail, obtained by adding a data logger to a meat container.

type of applications since they are easy to implement, e.g. adhered to the pallet, accurate and have a large data capacity. A secondary but important effect of the sudden appearance of data loggers on a pallet is that it will increase the quality awareness of people in the field.

Usually fixed readers are used but optional portable readers can be used which make the data easy accessible. Data are easily shared between partners since communication via the Internet is possible. However, robust data loggers having a large data capacity can be costly. If the reusable devices are expensive, they need to be returned to their owners, which is a logistic challenge in itself. For full-scale introduction of condition-measuring technology in reusable transport units (pallets or roll containers) data loggers or i-buttons can be used especially when the data can be transferred to a reader using RF.

In a future scenario, suppliers of sensitive food products may share their data collected with data loggers anonymously with their colleagues via an internet application. A third party gathers and analyses the information and, with a log service, all members can compare their chain performance with other players in the market. A practical database, set up by linking together all information, will in time show the effects of chain optimisations. First initiatives in this direction have already started (www.E-faqts.com).

10.7.4 Scenario D: Smart active tag on every package?

Market development of RFID tagging is rapid. Item level tagging for groceries is expected in a decade. New production technologies will decrease tag prices. Research projects are dedicated to replace silicon-based chip technology with sustainable organic alternatives. Other developments are focused on integrating sensors in the RFID tags. This will all lead to improved quality visibility in the food supply chain.

Consider a transport pallet containing a pile of food products. In a future scenario, each product packaging can contain an RFID label with battery and a sensor. The tag contains the identification and logistic information of a product. The sensor monitors a parameter that is related to the quality of the product. This can be temperature or, for instance, a volatile component related to the amount of bacteria in the package. The battery enables the tag to store monitored information (portable data file) and actively send the information to a reader. Passing the pallet through a gate containing a reader, all RF labels can be read simultaneously since no line of sight is needed (that is, after the current problems with metal and liquid shielding and electromagnetic interference occurring at some radio frequencies have been solved).

The reader-gate can be located at any point in the supply chain, for instance entering a distribution centre. The identity and origin of the product will automatically be recorded together with the temperature history or volatile concentration. The reader is connected to a computer or network, which contains keeping-quality models that can predict the quality of products based on the temperature to which they have been subjected. The computer can also contain models relating the volatile concentration in a package to the amount of bacteria present in the product.

The product identity tells the computer which model is applicable and, from the temperature history, the quality of the product may be calculated, or the number of bacteria present in each package may be calculated from the monitored volatile component concentration. Preferably, the models would estimate the remaining lifetime of the product, i.e. how long before the quality decay is such that the product cannot be sold. Based on this information, it can be decided whether these products may be stored for another few days or if they need to be transported to a retail shop as soon as possible?

This scenario is characterised by a high level of chain transparency. Important data travels physically with the product and is accessible to every chain partner at any moment. Data can also easily be shared via the Internet. A big issue will be to transfer data into useful information. To keep the tag costs affordable, the amount of temperature data logged on each package will probably be much less than in the case of data loggers. But a temperature profile of each package together with identification data of each product will be a huge assignment in data management.

Empirical models will be necessary to convert temperature data into quality parameters of food products. Also the relation between volatile component concentrations and amount of bacteria in a package has to be modelled. The models can be simple or extensive but either way will require a lot of experimental data. Alternatively a self-learning system can be developed in which historical data of temperature profiles and quality parameters are used to predict the quality of the product at hand.

Finally, the quality parameters and amount of bacteria have to be transferred into action steps. What to do with a product that enters a distribution centre with a bacteria count of 5×10^6 cfu g^{-1}? This information has no use for personnel at a distribution centre. They only want to know if the product has to be shipped to A or to B.

How realistic is the scenario as illustrated here? Profit margins on food are not very large and costs for this kind of technology will have to come down to make it viable. On the other hand, regulations for tracking and tracing are constantly elaborated and food safety is of increasing importance. The technology described here allows monitoring of food quality on a consumption unit scale. The first applications for item level tagging can be expected for blood transfusion products and medicines. But soldiers can also benefit from this type of technology when their food products distributed in the uncontrolled logistic chains as occurring in military operations are provided with smart active tags.

10.7.5 Scenario E: Genomics predicts quality decay

Decades from now, apples might still grow on trees. Imagine that a robot comes by and samples an apple to test its gene expression. Based on this information as well as an image analysis of the size, colour and shape of the apples, the robot decides it is time to harvest the apples. An active biotag in the form of an eatable stamp containing DNA markers is placed on the apples. All information is stored and sent wirelessly to a central computer. The DNA marker in the stamp serves to identify the apples and, in the computer, it is linked with all the relevant information on type, origin and growing conditions. The gene expression at the moment of harvest gives information on the product condition and its sensitivity to temperature abuse in the supply chain. Gene expression of the apples is analysed to decide on the optimum supply chain and chain conditions. This determines the logistic path and when and were the apples will be available for consumption.

Is this fantasy or simply a matter of time before it can be realised? In theory, it is possible and a lot of research money is being spent to make it happen. However, biological variation may limit the applicability of the technology. To overcome this, the sampling size has to be enlarged. Therefore, in the future, the final question will be whether the costs will exceed the benefits.

10.8   Conclusions

For quality-oriented traceability a lot of technology is available. Trends in technology development are towards smaller and cheaper devices and more detection of parameters that are directly related to quality. The challenge is to put the type of technology in place that best fits the goals set by quality management.

10.9   References

Agriholland (2003). Varkensvlees via DNA terug te leiden naar de producent. (http://www.agriholland.nl/nieuws/artikel.html?id=35261)

AIM (2002). Global Yearbook & Buyer's Guide by AIM (The Association for automatic Identification and data capture technologies).

Alocilja, E C and Radke, S M (2003). Market analysis of biosensors for food safety, *Biosensors and Bioelectronics* **18**, 841–846

Jorma, F *et al.* (2003). RFID spoilage sensor for packaged food and drugs, WO03044521 (patent)

Smolander, M, Hurme, E and Ahvenainen, R (1997). Leak indicators for modified atmosphere packages, *Trends in Food Science & Technology*, April 1997

Urlings, B (2002). Biotagging van Boer tot Consument, presentation at workshop Tracking & Tracing, Wageningen, March 21, 2002.

Vernède, R, Verdenius, F and Broeze, J (2004). Traceability in food processing chains. State of the art and future developments. KLICT position paper.

Wilson, T P and Clarke, W R (1998). Food safety and traceability in the agricultural supply chain: using the internet to deliver traceability. Supply Chain Management Volume 3, Number 3, pp. 127–133.

11

Data carriers for traceability

A. Furness, AIM UK

11.1 Introduction

To achieve traceability of foods within supply chains it is essential to identify the food items concerned and to provide a seamless facility for maintaining identification of those items from source to consumer. *Primary* and *secondary* aspects of identification are necessary to ensure traceability (Fig. 11.1). Primary identification is about identifying the source components whether they are crop-based, fish or animals. Various techniques are available for identification at this level. These are primarily DNA or other

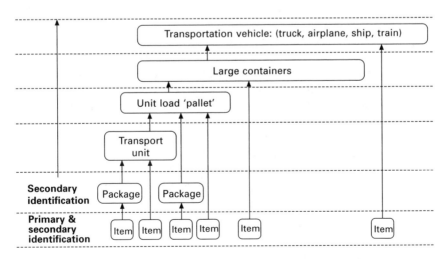

Fig. 11.1 Primary and secondary identification in traceability.

molecular- based analytical methods that can be used to identify an individual entity to a reasonable degree of statistical confidence and they invariably feature extraction techniques. The most significant aspect of these techniques is that they identify the entity by measuring a unique component or characteristic.

In contrast, secondary identification is based upon attributing an identifier to an entity in a form that can be attached or conveniently accompany that entity through or partly through the supply chain. Where such identifiers are used only partly within the chain, other identifiers must be introduced to accompany the entities, part-entities or combined entities as they are processed or handled within the chain. These secondary identifiers and their associated carriers are the subject of this chapter.

Secondary identification relies upon the use of numbers or alphanumeric strings where possible exploiting available standards for numbering and identification. The EAN.UCC system standard is significant in this respect. EAN stands for the International Numbering Association (formerly the European Article Numbering Association) and UCC, the Uniform Code Council (based in the United States) who have effectively joined forces to develop and promote international standards for numbering and identification.

Secondary identification can be used at source, to tag animals, for example, and may also be used to identify batches of raw materials such as cereal crops and genetically modified organisms (GMOs). By far the greater use of secondary identification is to identify entities at various levels of packaging from single items, through packages, transport units, unit loads at pallet level, containers and transport units. It is important at this point to qualify what is meant by an item. In one context, it is considered to be the lowest, discrete level at which a foodstuff or other entity can be distinguished for handling purposes. In more general terms, an item may be considered to be any identifiable entity. When referring to item management, for example the more general view is adopted. The context in which it is used generally conveys its meaning.

In order to use a secondary identifier for item management and traceability purposes a data carrier is required. This is the physical entity which is attached to, directly marked upon or accompanies the item and carries in some machine-readable form the identification number of alphanumeric means of identification. Identifiers can allow either human or machine readability, or both. For the automated systems approach to achieving traceability, machine-readable data carriers are required. A range of data carrier technologies and an even wider range of products and systems are available to support identification at these various levels. The technologies that are considered particularly relevant to the needs for food traceability include:

- linear barcodes
- two-dimensional codes (multi-row bar, matrix and composite codes), including direct marking
- contact memory devices

- radiofrequency identification (RFID)
- smart labels (passive and active devices)

For open systems usage, which is invariably the case for food supply chains, it is essential that the identification codes and any additional information relating to the item, such as batch number, weight, volume, other identification numbers, use or sell-by-date, adhere to a particular identification standard. Through standardisation, the hierarchy of packaging and item traceability can be better achieved. The EAN.UCC standard not only provides the necessary coding structures for identification of items, and other entities such as location, but also specifies adopted data carriers, presently confined to bar and composite codes but currently pursuing standardisation for radiofrequency identification (RFID)-based data carriers.

Automatic identification and data capture (AIDC) is a term that encapsulates the requirements for achieving traceability, but is also a term that denotes both an industry that serves a growing user community and a range of technologies and associated principles for automated item-level data acquisition. The industry representative body that deals with AIDC support is the Association for Automatic Identification Manufacturers (AIM) for which there is a global body and a number of AIM affiliates worldwide, including AIM UK and AIM Europe. The central AIM website is www.aimglobal.org at which a wealth of information is available on technologies and standards.

In their most common form, AIDC technologies and associated systems automate the capture of item-based data. This item-based data can generally be structured to generate an individual- or class- or type-based identification. These identifiers can be used for process-based identification purposes, but also as a unique key to retrieve a record for the item from a local or distributed database. Where data is used in this way, it is commonly termed 'licence plate' application. The identifier under such circumstances may have no physical meaning, being simply a unique numeric or alphanumeric key to remotely held data. A driver's licence number or social security number uniquely identifies an individual whereas a supermarket barcode uniquely identifies a particular product type but not each instance of the class because all identical products carry the same identifier. In addition to licence plate applications, many AIDC data carrier technologies can carry significant data payloads and can be configured to act as local item-based data caches that carry or relay data between distributed heterogeneous and distributed systems. This approach can be extremely powerful where a guaranteed timely connection to a remote data source is impracticable, impossible or uneconomic.

11.2   Linear barcode systems and EAN.UCC adopted symbologies

Linear barcode data carriers are the most prominent and well established of the AIDC data carrier technologies having been in widespread use since the early 1970s. They are a familiar sight on products in retail stores and on a wide variety of packaging and containers. They are used widely in manufacturing, asset management, document tracking, access control, warehouse management and distribution.

At first sight, one barcode may look much like any other. Although simple in concept the way in which the bars and spaces are structured to carry data in digital form (the 1s and 0s, representing the data encoded) can be somewhat sophisticated. The rules by which they are structured determine to a large extent the type of barcode and the attributes they exhibit. The data carrying part of the barcode symbol comprises a number of alternating dark (bar) and light (space) rectangular elements of variable width (some are based upon narrow and wide bar/space elements). In addition to the bars and spaces that are used to represent the encoded data other structural features of the symbol can be distinguished that facilitate the reading or scanning process and enhance the security or integrity of the symbol.

Typically, a barcode symbol may incorporate some or all of the following additional features (Fig. 11.2)

- **Quiet zones** (start/stop margin) – relatively broad spaces or zones on both sides of the bar/space structure. These zones provide a reflectance reference region similar to that of the space elements, at either end of the barcode, to allow an optical reader to establish a start condition or reference level for reading the code.
- **Start and stop characters** – particular groupings of bars and spaces, incorporated into the symbol to distinguish the start and the end of the encoded data. A symbology that allows a symbol to be read in either direction is said to be bi-directional.
- **Check character(s)** – a check digit or character derived with reference to the encoded data to provide a check on decoding if the data has been received without error. Depending upon the symbology, the check features may be mandatory or an optional feature. Human readable characters corresponding to the data content are often printed below the bars and

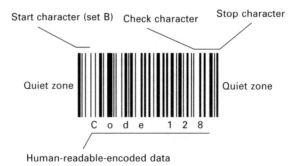

Fig. 11.2 Barcode features.

spaces of a barcode symbol to allow manual keying should a problem arise that denies a read in the automatic capture process.

- **Self-checking structures** – bar/space encoding structures for the character sets designed to avoid valid reads being accepted as a result of a single printing defect causing an elemental transposition error in the symbol.
- **Inter-character gaps** – a feature of some of the earlier symbologies wherein gaps between encoded characters were provided, ostensibly to facilitate printing of sequential symbols using mechanical, numbering wheel devices in a letterpress printing process.
- **Vertical redundancy** – the height of the bar/space symbol which allows different scan paths through the symbol to be tried in the event that the scanning attempt fails in the path due to localised voids or specks in the barcode structure.

Reading these symbols typically involves directing a beam of red light across them and detecting the changes in the amount of light reflected from the surface. These variations are converted into a digital signal from which measurements are derived electronically to determine, through a decoding process, the characters that the signals represent.

The rules determining the way in which the bars and spaces represent characters and numbers is known as a symbology. Symbologies of various kinds have been developed to accommodate a wide range of data requirements. Many of them have been standardised in the form of standard symbology specifications, making them available for open systems applications. Some symbologies have been developed to accommodate numerical data only, some to accommodate alphanumeric data, some to enhanced data density, some to accommodate the use of inexpensive printing methods and others to exhibit higher security of data and reliability in reading. The differences between the various symbologies lie mainly in the range of characters that they can encode, and the way in which these individual characters are represented by patterns of bars and spaces.

Many symbologies have been developed over the years, fortunately the majority of applications requiring linear barcodes are satisfied by just a few of these symbologies, generally supported by standard symbology specifications. The more popular of the symbologies include:

11.2.1   Interleaved two of five symbology

Interleaved two of five is a widely used, relatively high density, numeric-only symbology. It is a narrow–wide code in which characters are represented by five elements, two being wide. Data is encoded, as its name suggests, as pairs of characters, the first comprising five bars and the second by five spaces interleaved between the bars (Fig. 11.3). Because of the pairing symbols require an even number of characters, but a lead zero may be used to accommodate odd numbers of characters.

Fig. 11.3   Interleaved 5 of 5 symbology.

Fig. 11.4   Bearer bar for interleaved 2 of 5 symbology.

Owing to the simple structure of the start and stop characters of the interleaved two of five symbology, parts of valid characters can be misread if the scan crosses obliquely across and off the symbol before reaching the end. The result under such circumstances is termed a short scan, delivering what would appear to be a valid set of digits. To guard against this occurring it is generally recommended that applications using interleaved two of five symbols use a fixed length format and the readers be programmed to accept valid reads that satisfy the length constraint. Short data messages can be padded with leading zeros. A further facility for guarding against partial scans yielding apparently valid, but erroneous reads, is the use of bearer bars. These are wide bars placed along the top and bottom of the barcode symbol. A partial scan is detected by the signal change arising from the characteristically wide bearer bar (Fig. 11.4).

To improve data security it is usual to incorporate a 'modulo-10' check digit in the final character position. The presence of the check digit may suggest that it can be used to detect errors due to partial scans. However, it cannot, as a single measure, solve this problem, hence the need for message length specification and bearer bars. Interleaved two of five is invariably used for industrial applications and for the outer-case of consumer goods, particularly where adopted using EAN format.

11.2.2   Code 39 or Code 3 of 9

Code 39 is one of the most widely used symbologies for industrial and non-retail distribution applications. It is specified in a number of industry application standards. Code 39 is a discrete, alphanumeric, narrow/wide symbology in which each character is defined by 9 elements, 3 of which are wide. The character set comprises 43 characters, the digits 0 to 9, alphabet A to Z and seven special characters (–.* (space) $/%). The full ASCII set may be encoded

Fig. 11.5   Code 39 symbology.

using pairs of characters, but yields an inefficient encoding density, poor in comparison with other symbologies, such as Code 128, that accommodates full ASCII (Fig. 11.5).

11.2.3   Code 128

Code 128 is an alphanumeric multi-width symbology capable of efficiently encoding the full ASCII character set (Fig. 11.6). Each character in code 128 comprises 11 modules (narrowest width elements) deployed in 3 bars and 3 spaces.[1] Three data sets are defined to accommodate the full set of 128 ASCII characters, each set selectable by use of an appropriate start character. Two of the sets contain the capital alphabet and numeric digits together with lower case alphabet in one set (B) and control characters in the other (A). The third set (C) uses the codeword symbols to represent paired digits, 00 to 99, thus providing a high-density (up to 24 digits per inch) numeric encodation set. Code 128 is gaining in popularity for a wide variety of applications.

11.2.4   EAN.UCC adopted symbologies

A number of symbologies have been adopted by EAN International and the Uniform Code Council to support the EAN.UCC system of numbering and identifiers. These symbologies are usually distinguished by an EAN prefix when naming them. There are also other distinguishing features that allow systems using them to identify them as EAN.UCC adopted symbologies.

The EAN.UCC specifications account for the largest single usage of linear

Fig. 11.6   Code 128 symbology.

---

[1] Code 128 is an example of a n, k code in which n denotes the number of modules to represent each character or codeword and k the number of bars and spaces in which the n modules are deployed. The maximum number of modules for any one bar or space is also often quoted (often denoted m). For Code 128, m = 4.

barcodes, not simply on consumer unit packaging but also on transit packaging and pallets at higher levels in supply chain applications. Because information requirements at different stages differ, the EAN-128 standard uses the special data formats distinguished as 'application identifiers' for defining the nature of application data, such as a batch number, a 'best before' date, order number and so on.

11.2.5   EAN-13, EAN-8 and UPC
The majority of retail items to be found in shops and supermarkets carry what are known as the EAN-13 (13-digit) or EAN-8 (8-digit) types of barcode symbol, structured in accordance with the EAN.UCC symbology specifications. EAN-8 symbols are used for in-store small item marking. While the EAN-13 and EAN-8 symbols exhibit some of the general features so far described for barcode symbols, they also exhibit structural features that are peculiar to this symbology. A similar form to EAN-13 and EAN-8, used in the USA and Canada, encodes 12 digits and is known as the UPC-12 symbol.

The EAN-13 symbol is an example of a continuous, numeric only, fixed-length code of 13 characters, 12 of which are in the form of bars and spaces and one, leading character, is derived from the way in which the first six characters are structured. By continuous it is meant that there are no inter-character gaps thus allowing more efficient use of space (Fig. 11.7). One of the key features of the EAN-13 and EAN-8 symbols is the ability to read them using omnidirectional scanning. They are effectively two symbols in one – left and right, separated by the centre bar-space structure.

The symbol is an example of a particular group of multi-width barcodes. A minimum bar width is defined in such codes, known as the module width or X dimension. Each character within an EAN-13 symbol comprises seven of these modules deployed as two bars and two spaces. The symbology accommodates digits 0 to 9, start, stop and centre features represented in each of three sets, A, B and C. Digits to the right of the centre bars are structured in accordance with set C and those on the left from sets A and B. The order with which the digits on the left are structured using sets A and B determines the leading digit, which is not itself represented in bars and spaces. The UPC-12 symbol uses one set to encode the 12 digits.

Fig. 11.7   EAN-13 symbology.

11.2.6   EAN Interleaved two of five

The International Numbering Association (EAN) have adopted Interleaved two of five (ITF for short) as the symbology for coding transit packaging, partly due to the ease with which the symbols can be printed onto packaging materials such as corrugated cardboard. As the EAN form of ITF accommodates 14 digits, it is often referred to as ITF-14.

11.2.7   EAN 128

EAN has also adopted Code 128 symbology as a means of carrying EAN.UCC numbers and specifying supplementary data, such as batch numbers, for product identification purposes. To avoid confusion with other Code 128 labels that may be included on a package or product for other purposes, a function character, reserved for EAN, is included in symbols immediately after the start character (Fig. 11.8).

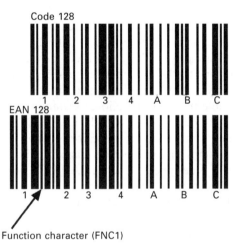

Fig. 11.8   EAN Code 128 symbology compared to standard Code 128 symbology.

11.2.8   UCC Reduced Space Symbology

Reduced Space Symbology™ (RSS) is a family of three linear symbologies and variants (seven in total) specifically developed to accommodate the EAN.UCC Global Trade Item Number (GTIN) on space-constrained items, where existing linear symbologies could not be used. The RSS symbols can be linked to three varieties of two-dimensional symbologies to form what are designated Composite Components™. Both the RSS and Composite symbologies are supported by AIM International Symbology Specifications. Although UCC hold the patents for these symbologies, they have placed them in the public domain to allow users to freely develop commercial applications.

Fig. 11.9 UCC Reduced Space Symbologies (RSS). (a) truncated and (b) stacked (note 1 module high row).

The RSS structures have been designed to satisfy the emerging market needs for more information to be included in barcode symbols and to provide a robust symbology for supporting barcode symbols on smaller item labels (Fig. 11.9). The three basic RSS structures are summarised as follows:

- **RSS-14**™ is a mixed multi-width symbology distinguishing 'inner' and 'outer' character structures capable of efficiently encoding 14 digits of numerical data that can be used to carry a Global Trade Item Number (GTIN). Further data can be accommodated by linking to a Composite Component™ that can contain up to 2361 data characters. Truncated and stacked formats for the symbology specified, the former to support higher density labelling and the latter to reduce the overall length of a symbol so it can better satisfy particular packaging configurations. A stacked format structure is suitable for omni-directional scanning.
- **RSS Limited**™ is designed to encode EAN.UCC item identification numbers together with packaging indicators of 0 or 1 in a linear symbol for use on small items that will not be scanned at point-of-sale.
- **RSS Expanded**™ is designed to encode EAN.UCC primary identification numbers and supplementary data strings such as item weight and 'best-before-date'. The symbols structure is designed for omni-directional scanning. A stacked version of RSS Expanded can also be distinguished to accommodate requirements where the normal symbol would be too wide.

Any of the RSS family members can be printed either as stand-alone linear barcode symbols or as composite symbols using specified EAN.UCC 2D Composite Components printed directly above the RSS linear component. A linkage flag is used within the RSS symbol to denote the presence of a 2D composite component.

11.2.9 Composite symbologies

Composite symbologies comprise a family of structures in which an EAN.UCC linear symbol is linked to a 2D symbol. The composite components (CC) support supplementary application identifier data for the linear EAN.UCC component. Three versions of the CC are specified and designated CC-A, CC-B and CC-C, to accommodate different symbol sizes and data capacity.

Fig. 11.10   A composite symbol.

CC-A can encode up to 56 alphanumeric characters, CC-B up to 338 alphanumeric characters and CC-C up to 2361 alphanumeric characters. The CC versions A and B can be linked with RSS, EAN/UPC and UCC/EAN-128 symbols. The CC-C version is used with UCC/EAN-128 symbols (Fig. 11.10).

Composites are generally intended for applications where different parts of the information may be required at different stages or aspects of the item supply line and in cases where there are restrictions on the amount of space available in which, for example, a second linear barcode could be placed. Certain users of these codes will need only to read the item identification in the linear symbol; others may need the full supplementary information carried in the two dimensional composite component. For further information on the EAN.UCC system and adopted data carriers visit their website at www.ean-ucc.org.[2]

11.2.10   Barcode labels

An important and popular use of barcode symbols is in printed form on labels. On retail food products the barcode symbol is invariably part of the item label or item packaging and is usually printed as part of the label artwork. On non-retail items, barcodes may be applied by direct marking techniques onto packaging or carton substrates, usually by ink jet printing.

Discrete labels are also possible. A wide variety of label stocks are available to satisfy an equally wide variety of application needs, particularly in respect of environmental and handling demands. Some applications require labels that are scratch resistant and able to withstand high humidity for example. A label may comprise several layers, each requiring consideration in given applications. For example, the substrate and its ability to take a printed symbol without distortion or spreading of ink, or the strength of adhesion of the adhesive layer to meet attachment and detachment requirements.

For producing barcode labels a range of printing techniques and associated products are available including:

* **Ink jet** – low-cost print facility, but care has to be taken to ensure readable codes by avoiding ink spread and distorted bar edges. Often suitable for outer-package marking and potential for direct marking onto food items using suitably safe inks.

---

[2] GS1, a voluntary standards organization is charged with the management of the EAN.UCC System and the Global Standard Management Process (GSMP).

- **Laser** – medium-cost print facility offering quality symbols providing due care is taken with respect to toner level and paper stock used.
- **Direct thermal** – higher-cost, good quality print capability, but requiring special coated-paper stock. Images on such paper fade over a period of up to three months, although label stock (full barrier stock) is available that allows images to be retained for up to two years.
- **Thermal transfer** – higher-cost, good quality print capability, avoiding the need for special coated paper. Non-paper, filmic stock may also be used to provide long life, durable labels capable of withstanding significant abrasion and harsh environments.

These options make barcode printing and reading very economical, flexible and versatile, and enable the technology to be used in a wide range of applications in widely differing situations. Readily available software and printing products mean that barcode labels are generally being produced in-house to meet on-demand applications. This is particularly significant in food supply chain applications. However, there are occasions where it may be appropriate to buy in or obtain what are called barcode master symbols for use in printing artwork. The latter are invariably appropriate where a large number of the same barcode symbol is required. A decision to buy in may also arise where sequentially numbered barcode labels are required on particular label stock. The quality of labels and the constituent symbols are an important aspect of effectively applying barcode technology and verification devices should be used as appropriate to provide quality assurance (Table 11.1).

11.3   Two-dimensional coding

An extension of the linear (1D) barcode is the two-dimensional (2D), in the form of either a multi-row or matrix symbol. Such symbols can contain large data payloads in excess of 1500 characters in some cases and thus offer data-carrying options that can easily exceed the carrying capacity of linear barcodes and satisfy complementary application needs. Such structures may also allow greater density capability than linear barcodes and thus offer prospects for labelling small items that would be difficult or impossible to achieve with linear barcodes. Through appreciating the attributes of both categories of 2D data carriers a better foundation for choosing technologies appropriate to needs can be established.

11.3.1   Multi-row barcodes

Multi-row (or stacked) barcodes may be viewed as a series of linear barcode symbols stacked directly, one on top of the other. However, they are somewhat more sophisticated in structural terms than linear barcodes. The row height

Table 11.1 Summary features for linear barcodes and system components

Data attributes	Low capacity, typically 15 to 50 characters, carrying capability, depending upon symbology and the symbol form, favouring 'licence plate' usage (code to locate data stored elsewhere). Symbologies available to accommodate numeric, alphanumeric, ASCII and other character sets and various encoding densities.
Carrier form	Variety of symbol (barcode) forming techniques, primarily printing methods for barcode labels and documents but also pierced metal, impressed and composite formed symbols on a range of substrates. Variety of labels and other substrate forms and symbol realisations to suit a variety of applications and user environments.
Standards	A range of symbologies supported by Uniform Symbology Specifications and ISO equivalent standards, with a number of area-specific, open systems, applications adopting particular symbologies.
Costs	Low-cost, label-based symbol formation by a variety of printing and other techniques. Wide range of symbol formation software, printer hardware, label products, scanning systems (portable and fixed-position) and verifiers for assessing symbol quality.
Product reading devices	**Wand readers** – low-cost, entry-level technology, popular for satisfying low-volume, close proximity application needs. **Laser scanners** – wide range of scanners to satisfy a range of application needs in respect of auto-discrimination of different symbologies, range, depth of field, speed and reliability of read and form factors to satisfy various environmental needs. **CCD scanners** – range of scanners to satisfy a range of application needs, width of read head determining the size of barcode symbol that can be accommodated. Typically lower cost than laser scanners for a given performance specification, but limited on range and depth of field – essentially close proximity reading devices.
Verifiers	A variety of linear bar code verifiers are available offering various levels of measurements and reporting facilities. Important to ensure compliance with appropriate ANSI verification standards.
Reader interfaces	**Keyboard wedge** – facility to connect a barcode reader decoder output between a keyboard and computer host so that data so derived appears as if entered via the keyboard. The keyboard continues to operate normally, so care has to be exercised in distinguishing and handling reader entered data. **Wand emulation** – non-decoded output interface in which the signal effectively emulates the output from a wand reader without a decode facility. Some form of decoder is then required to interface to the host computer or data management system. **RS232** – a standard serial interface often incorporated into bar-code readers allowing easy and effective interface to PCs. **Universal Serial Bus (USB)** – an increasingly popular form of serial interface for connecting readers to currently available computers equipped with USB ports. **Wireless interfaces** – various wireless communication systems available for wireless connection between readers and host data handling systems.

Table 11.1   Continued

Label software	A wide range of label and data management software is available with correspondingly wide ranging prices. It is important to ensure software meets user needs in respect of suitability for application needs including compatibility with existing software or data management system, quality and compliance of symbols with existing standards. Barcode TrueType fonts are now available for Microsoft Windows applications. While these provide a convenient and easy way of incorporating barcodes into documents and labels care has to be exercised in using them directly in document and spreadsheet applications. Some symbologies, because of technical issues concerning representation in font form characters, require particular consideration in the way they are handled. Windows-based label packages generally accommodate these requirements.
Printers	Wide range of printers available to meet wide-ranging application needs.

in a multi-row barcode is generally lower than that of a linear barcode, in order to increase the spatial efficiency of the symbol. Because of this and the greater data density there is a greater risk of data loss if damage to the symbol occurs. To accommodate this problem, most of the symbologies, but not all, incorporate powerful mathematical error detection and correction techniques. At the cost of increasing the number of symbol characters in the symbol, these techniques ensure that a number of 'erasures', or undecoded characters, and 'errors', which are mis-decoded characters, can be identified and corrected. Some symbologies offer fixed levels of error correction; others allow the user to select a level depending on the perceived risk of symbol damage in an application. In addition to these features, multi-row symbologies also generally include some means of indicating the total number of rows in the symbol and the position of each individual row in the sequence.

The early multi-row barcodes had maximum capacities in the order of 100 characters, more recent symbologies allow up to 2000 or so characters to be encoded in a single symbol (higher for numeric compaction) and the option for concatenation or linking of symbols. The lower capacity symbologies generally enabled short messages to be incorporated into a smaller area than linear barcodes and were adopted in, for example, the pharmaceutical and electronics industries, both of which need to mark small items. The potential may thus be seen for labelling small food items. The more recent higher capacity symbologies are intended more for the 'portable data file' type of application, in which stand-alone information needs to travel with the item where, for example, it is not possible to readily access a central database. Here the prospect may be seen for including additional machine-readable information on food items.

A number of symbologies have been standardised as ISO and AIM Uniform Symbology Specifications. Those that may be seen to have particular significance in food traceability include:

Fig. 11.11   A PDF417 symbol: 36 characters, approx. ×10 magnification, level 5
security (actual size determined by print and read requirements).

- **PDF417** – First of the high-capacity two-dimensional symbologies, it is designed to encode numeric, alphanumeric, full ASCII, extended ASCII and byte data, and access capability to other data interpretations and character sets (Fig. 11.11). Symbols from 3–90 rows and from 1–30 columns can be constructed. The maximum data characters per symbol are 1850 alphanumeric, 2710 digits and 1108 bytes depending upon the compaction mode selected. It incorporates powerful error detection and correction capabilities and allows user selection of nine levels of error-correction capability. PDF417 has been specified as the symbology for carrying shipping data in ISO15394 and for portable data files in a number of industry association applications, including AIAG (Automotive Industry Action Group) and EIA (Electronic Industries Association of America).
- **Micro PDF417** – MicroPDF417 incorporates the standard PDF417 encoding but with modified start/stop and row identification features to provide a more compact structure. It is designed for applications with a requirement for improved area efficiency but without the need for PDF417's maximum data capacity. A limited set of symbol sizes are specified together with a fixed level of error correction for each symbol size. The symbology is suitable for small-item marking and has been selected as one of the 2D Composite Components for EAN/UCC Composite symbols.

Printing techniques for these symbols are much the same as those for linear barcodes, care being necessary to ensure the appropriate resolution can be achieved for the symbol needs in a given application. Reading, however, requires the data in each row to be read and the complete message reconstructed in the correct sequence before it is output from the reader to the host computer.

Multi-row symbologies were designed, generally speaking, to allow the use of laser-based barcode scanning technology to be applied for the purpose of reading them. However, conventional linear barcode scan engines are not appropriate for multi-row barcodes. Laser scanners in which the scanning beam is moved up and down the symbol, as well as scanning horizontally (known as raster scanners), are typically required for this purpose. An alternative reading approach, becoming increasingly common, uses image capture techniques, in which the complete symbol image is captured, analysed by image processing and decoded using appropriate, scanner resident, software (Table 11.2).

Table 11.2 Summary features for multi-row barcodes and system components

Data attributes	Low- to high-capacity character carrying capability, depending upon symbology and the symbol form, favouring portable data file usage. Symbologies available to accommodate numeric, alphanumeric, ASCII and other character sets and various encoding densities.
Carrier form	Essentially printed label and document carrier form. Variety of label stocks available to satisfy particular application requirements in respect of durability.
Standards	A range of symbologies supported by Uniform and International Symbology Specifications.
Costs	Low-cost, label-based symbol formation by a variety of printing and other techniques. Range of symbol formation software often now supported in linear barcode support packages, printer hardware, label products, scanning systems (often including capability to read the more popular linear barcode symbologies).
Product reading devices	**Laser raster scanners** – a range of scanners available generally including capability to auto-discriminate and read the more popular linear barcode symbologies. **CCD scanners** – range of scanners to accommodate the more popular 2D symbologies, both multi-row and matrix. Particular aspect ratios for multi-row symbols may present difficulties for CCD devices to read, depending also on the scanner head size.
Verifiers	No standards are presently available to support the verification of multi-row barcodes.
Reader interfaces	**RS232** – a standard serial interface often incorporated into barcode readers allowing easy and effective interface to PCs. **Universal Serial Bus (USB)** – an increasingly popular form of serial interface for connecting readers to currently available computers equipped with USB ports. **Wireless interfaces** – various wireless communication systems available for wireless connection between readers and host data handling systems.
Label software	A range of label and data management software is available to accommodate data from barcode readers (including multi-row derived if decoding is performed within the reader. It is important to ensure software meets user needs in respect of suitability for application needs including compatibility with existing software or data management system, quality and compliance of symbols with existing standards.
Printers	Wide range of printers available to meet wide-ranging application needs.

11.3.2 Matrix codes

Matrix codes are a further form of printable two-dimensional data carriers, similar in many respects to multi-row barcode data carriers. However, they differ in the way the data is held in spatial terms. Instead of bars and spaces the data encoded into matrix codes are stored in cells, which are filled or unfilled to represent the binary data, using a contrasting colour to distinguish the filled cells. The cells are arranged generally on a square grid, similar to a chessboard, although some symbologies use other geometric arrangements. In the majority of cases, the individual cells abut their neighbours directly, though there is a small subgroup, known as 'dot codes' where the individual cells do not touch. The data containing part of the cells are surrounded by a small area of clear space – making them particularly suitable for marking of items by punching of holes in the material.

Matrix codes are capable in theory of representing the highest amount of information in a given area for a given unit dimension. Consequently, they are often used for the marking of small components such as integrated circuits on which it is not uncommon to satisfy a need to encode fifty or so characters in two or three millimetres square. This would be impossible with conventional linear and even difficult with a multi-row barcode symbol. Such techniques open up the prospects for direct marking of foodstuffs providing sufficiently safe and cost-effective methods can be derived.

Because the fundamental code structure is binary, encoding data in matrix codes is a question of converting it to a bit stream and then arranging it according to the rules of the symbology. Normally, a light cell corresponds to a zero bit and a dark cell to a one bit. However, some of the symbologies can also be decoded if the dark and light values are reversed, as may be the case when the symbols are marked by laser etching processes on a dark or metallic surface.

For matrix codes, the use of robust error detection and correction techniques is essential. This is because a comparatively small defect could still render an individual cell unreadable, or transpose the apparent value of a bit from zero to one or vice-versa, thus leading to erroneous decoding. Typically with these symbols the amount of error correction used will enable up to between fifteen and twenty-five per cent of the symbol to be damaged whilst allowing the entire data message to be reconstructed correctly.

Matrix code symbols are generally characterised by features that allow the symbology to be recognised in a captured image (a finder pattern), allow its orientation or alignment to be established and the data to be retrieved. All the matrix codes are capable of representing the full ASCII and extended ASCII character sets. As with multi-row barcode symbologies a number of matrix symbologies have been developed some of which are now supported by AIM International Symbology Specifications, including:

- **Maxicode** – is a low-capacity matrix code symbology capable of encoding ASCII characters and other character sets through the ECI protocol. The Maxicode was developed by the United Parcel Service (UPS) to meet a requirement for identifying packages on fast conveyors (150 m min^{-1}) for sorting purposes. The symbols are of fixed dimension ($28 \text{ mm} \times 27 \text{ mm}$) and characterised by a centrally positioned *finder pattern* and hexagonal data elements (Fig. 11.12). The symbol can accommodate up to 138 digits or 93 alphanumeric characters. The symbol also includes powerful error

Fig. 11.12   A Maxicode symbol: note fixed dimension for symbols, 28 mm × 27 mm.

control facilities. Since it was designed for identification and sorting purposes its potential can be clearly seen in other areas of handling and tracking of packages and similar items.

- **DataMatrix** – is a high-capacity matrix code symbology capable of encoding ASCII characters and other character sets through the ECI protocol. Symbols may accommodate up to 3116 digits, 2335 alphanumeric characters or 1556 bytes of data. A wide range of symbol sizes can be distinguished, both square and rectangular, to meet data and label/marking requirements. DataMatrix is particularly suitable for small-item labelling and particularly direct marking of items, such as electronic components, silicon wafers and pharmaceutical unit dose containers, and is also being applied to food product identification, although, as yet, there are no EAN.UCC adopted matrix codes (Fig. 11.13).

- **Aztec Code** – is a high-capacity matrix code symbology capable of encoding ASCII characters and other character sets through the ECI protocol. Symbols may be 'compact' or 'full-range' variable size structures. Fixed size, 8-bit 'Aztec runes' can be distinguished. Compact symbols can encode up to 110 digits, 89 alphanumeric characters or 53 bytes, while full-range structures can encode up to 3832 digits, 3067 alphanumeric characters or 1914 bytes (Fig. 11.14).

- **QR Code** – is a high-capacity matrix code symbology capable of encoding ASCII characters, Latin, Kanji and Katakana characters and other character sets through the ECI protocol. A range of symbol sizes can be distinguished, the maximum data per symbol (largest version) can accommodate 7089 digits, 4296 alphanumeric characters, 2953 Latin/Katakana characters or 1817 (kanji) bytes. The symbology is suitable for high-speed sorting

 Example of a data matrix symbol: 18 characters × 30 magnification

(a)

 Example of a data matrix symbol at 10 × magnification

(b)

Fig. 11.13   A DataMatrix code symbol (a) 18 characters, × 30 magnification (b) 10× magnification.

Fig. 11.14   An Aztec code symbol: 18 characters, ×30 magnification.

Fig. 11.15   A QR code symbol: 18 characters ×30 magnification.

applications (QR standing for 'quick response'). It has also been used for data-carrier applications in manufacturing and for product ordering. Potential may also be seen for food item identification where fast response is required (Fig. 11.15).

A variety of other matrix code symbologies have been developed and supported by standard specifications including Code One (used for high-speed sorting, e.g. identifying plastic bottles together with information for recycling) and Ultracode (suitable for low-resolution, low-precision marking, high-capacity data-carrier applications and multilingual encoding). For further introductory information, see the AIM publication 'Understanding 2D Symbologies'. Matrix codes must be read by image-processing techniques, generally based on images captured with a two-dimensional array of CCD photosensors. Since it is important to know the position of each cell in both x and y co-ordinates, the use of linear scanning techniques is inappropriate (Table 11.3).

11.4   EAN.UCC numbering system

The EAN.UCC system provides an international alternative for plain language descriptions that greatly assist in the control and the speed with which goods are moved and associated information communicated. The system defines and promotes the following support features:

- The use of unambiguous numbers to identify goods, services, assets and locations on an open systems, worldwide basis. As such the system has

Table 11.3   Summary features for matrix code symbologies and system components

Data attributes	Medium- to high-capacity character carrying capability, depending upon symbology and the symbol form, favouring portable data file usage. Symbologies available to accommodate numeric, alphanumeric, ASCII and other character sets and various encoding densities.
Carrier form	Most symbologies suitable for printed label and document carrier form. Variety of label stocks available to satisfy particular application requirements in respect of durability. Some symbologies particularly suitable for direct marking on to a variety of substrates including metals, plastics, glass and other ceramics.
Standards	A range of symbologies supported by Uniform and International Symbology Specifications.
Costs	Low-cost, label-based symbol formation by a variety of printing and other techniques. Range of symbol formation software often now supported in linear barcode support packages and printer hardware.
Product reading devices	**CCD scanners** – range of scanners to accommodate the more popular 2D symbologies, both matrix and multi-row. No standards are presently available to support the verification of matrix code symbols.
Verifiers	**RS232** – a standard serial interface often incorporated into barcode readers allowing easy and effective interface to PCs. **Universal Serial Bus (USB)** – an increasingly popular form of serial interface for connecting readers to currently available computers equipped with USB ports.
Reader interfaces	**Wireless interfaces** – various wireless communication systems available for wireless connection between readers and host data handling systems.
Software	A range of label and data management software is available to accommodate data from barcode readers (including matrix-derived data if decoding is performed within the reader). It is important to ensure software meets user needs in respect of suitability for application needs including compatibility with existing software or data management system, quality and compliance of symbols with existing standards. Proprietary systems with supporting software available to support direct marking applications.
Printers	Wide range of printers available to meet wide-ranging application needs (see linear barcodes).
Direct marking systems	Laser engraving, peening and other direct marking systems are available with supporting software to support direct matrix code marking.

been designed to overcome the limitations of using company-, organisation- or sector-specific coding systems and provide a vehicle for more efficient, more responsive trading.

- Representation of supplementary information such as serial numbers, best-before-dates, batch numbers and measurements, through the use of application identifiers.
- EAN.UCC Standard barcode symbologies defining how the numbers and identifiers can be represented in barcode symbols. The prospect is also provided for exploiting the system numbering and identifiers in other AIDC data carriers.
- A set of message structures, collectively referred to as EANCOM® messages, for Electronic Data Interchange (EDI)

11.4.1 Numbering for identification

The EAN.UCC system was designed as a common standard link between companies operating at different points along national and international supply chains. It was quickly realised that a classifying system for identifying all products and their intrinsic details would be too cumbersome as well as impossible to administer, and that the best approach was to use an identifier in which specific companies could be uniquely identified. The company identifier, or company prefix, is an intrinsic feature of the EAN.UCC system numbering structures and is allocated by an EAN Numbering Organization (UCC in the United States), of which there are many throughout the world. In the UK, e-centre is the representative organisation. An EAN prefix of two or three digits, allocated by EAN International to the EAN Numbering Organization, can also be distinguished as part of some of the numbering structures, invariably preceding the EAN.UCC company prefix.

The EAN.UCC company prefix, once allocated, provide access to applications using EAN.UCC identification standards, including, for example, those involving the identification of items, logistical units, returnable containers, locations and services. The company prefix is an integral part of each of the numbering structures identified within the EAN.UCC standards, with the exception of the eight digit, EAN/UCC-8, structures. For situations in which a company changes legal status as a result of mergers, acquisitions, partial purchases, splits and spin-offs EAN.UCC have produced guidelines to assist in accommodating or transferring numbering support.

Each company that joins the EAN system is allocated a block of numbers comprising an EAN and EAN.UCC company prefix and a range of numbers that the company can use to identify any of its products, locations or other business entities. The numbers themselves contain no information about the item or entity concerned, but provide a key to item or entity information held on a database or databases.

The EAN.UCC system defines a number of standard numbering structures appropriate to different applications. Each application determines how the number is to be used, but the standard requires that the numbers be used in their entirety and not as component parts. By complying with this requirement, standard use and worldwide acceptance of the numbering structure for the applications concerned is preserved.

Six Standard Numbering Structures presently comprise the EAN.UCC System:

- Global Trade Item Number (GTIN)
- Global Location Number (GLN)
- Serial Shipping Container Code (SSCC)
- Global Returnable Asset Identifier (GRAI)
- Global Individual Asset Identifier (GIAI)
- Global Service Relation Number (GSRN)

11.4.2 Global Trade Item Number (GTIN)

The Global Trade Item Number, formerly known as the Article Number, is used to uniquely identify trade items, with acceptability as a global standard. The trade item is recognised as any entity, product or service, for which there is a need to retrieve pre-defined item-attendant data at any point within a supply chain. The definition embraces both individual items and accommodated items in different forms of packaging.

The GTIN is defined as a 14-digit number (see Table 11.4) and, by suitable truncation, is used to define a family of four unique numbering structures as indicated below. In more technical terms, the truncation defining each of the other three structures is a right justification in a 14-digit field (Tables 11.5–11.7). The indicator allows each user to increase the numbering capacity when seeking to identify similar trade units accommodated in different packaging configurations. The user assigns the reference aspect of the number, the rules for allocation depending upon the application. The check digit is incorporated to provide a degree of error control, calculated in accordance with specified rules, using the digits being encoded.

Table 11.4 EAN/UCC-14 structure

Indicator	EAN.UCC identification of the items contained (without check digit)	Check digit
N_1	$N_2\ N_3\ N_4\ N_5\ N_6\ N_7\ N_8\ N_9\ N_{10}\ N_{11}\ N_{12}$	N_{14}

Table 11.5 EAN/UCC-13 structure

EAN.UCC company prefix and item reference	Check digit
$\longrightarrow \longleftarrow$	
$N_1\ N_2\ N_3\ N_4\ N_5\ N_6\ N_7\ N_8\ N_9\ N_{10}\ N_{11}\ N_{12}$	N_{13}

Table 11.6   UCC-12 structure

UCC company prefix and item reference	Check digit
$N_1\,N_2\,N_3\,N_4\,N_5\,N_6\,N_7\,N_8\,N_9\,N_{10}\,N_{11}$	N_{12}

Table 11.7   EAN/UCC-8 structure

EAN/UCC prefix and item reference	Check digit
$N_1\,N_2\,N_3\,N_4\,N_5\,N_6\,N_7$	N_8

Table 11.8   Serial Shipping Container Code (SSCC) structure

Extension digit	EAN.UCC company prefix and item reference	Check digit
N_1	$N_2\,N_3\,N_4\,N_5\,N_6\,N_7\,N_8\,N_9\,N_{10}\,N_{11}\,N_{12}\,N_{13}\,N_{14}\,N_{15}\,N_{16}\,N_{17}$	N_{18}

The numbers so distinguished provide unique identification when processed in a 14-digit data field, zeros being used to fill the leading (left) field positions as appropriate for the EAN/UCC-13, UCC-12 and EAN/UCC-8 structures. It is the field format for GTIN used in all business transactions and EDI messaging supporting the standard.

11.4.3   Serial Shipping Container Code (SSCC)

The Serial Shipping Container Code (Table 11.8) is used to identify logistical units – shipping containers or transport units and is eighteen digits in length to accommodate larger item reference numbers. The extension digit effectively increases the capacity of the SSCC and is used to qualify the application of the code by assigning values 0–9 in the data field according to specified rules.

11.4.4   Global Location Number (GLN)

The Global Location Number is used to identify legal entities such as companies, subsidiaries or divisions such as customers, suppliers, banks and so forth, functional entities such as specific departments within a legal entity, and physical location entities such as rooms, gates and delivery points. Like item-based identification numbers, the location numbers are keys to information stored elsewhere, giving details in respect of the location concerned. Thus, a simple number may provide access to a full address and contact details (Table 11.9).

Table 11.9   Global Location Number (GLN) structure

EAN.UCC company prefix and location reference	Check digit
$N_1\ N_2\ N_3\ N_4\ N_5\ N_6\ N_7\ N_8\ N_9\ N_{10}\ N_{11}\ N_{12}$	N_{13}

Table 11.10   Global Returnable Asset Identifier (GRAI) structure

Indicator	EAN.UCC company prefix and asset type	Check Digit
N_1	$N_2\ N_3\ N_4\ N_5\ N_6\ N_7\ N_8\ N_9\ N_{10}\ N_{11}\ N_{12}$	N_{14}

11.4.5   Global Returnable Asset Identifier (GRAI)

The Global Returnable Asset Identifier is employed to identify reusable entities such as containers and totes, that are normally used for the transportation and storage of goods. The structure is essentially 14 digits in length and accommodates the EAN/UCC-13 structure and an optional facility for adding a serial number up to 16 digits in length (Table 11.10).

11.4.6   Global Individual Asset Identifier (GIAI)

The Global Individual Asset Identifier is used to identify uniquely an entity that is part of an inventory within a given company. The number accommodates the EAN.UCC company prefix, together with a variable length, individual asset reference. This and the prefix are up to 30 digits in length (Table 11.11).

11.4.7   Global Service Relation Number (GSRN)

The Global Service Relation Number is an 18-digit code (Table 11.12) comprising an EAN.UCC company prefix and a service reference used to identify the recipient of services from a service provider. It is not therefore

Table 11.11   Global Individual Asset Identifier (GIAI) structure

EAN.UCC company prefix and individual asset reference
$N_1 \ldots N_I + X_{I+1} \ldots X_j$ where j is less than or equal to 30

Table 11.12   Global Service Relation Number (GSRN) structure

EAN.UCC company prefix and service reference	Check digit
$N_1\ N_2\ N_3\ N_4\ N_5\ N_6\ N_7\ N_8\ N_9\ N_{10}\ N_{11}\ N_{12}\ N_{13}\ N_{14}\ N_{15}\ N_{16}\ N_{17}$	N_{18}

identifying a particular person or legal entity but a relationship or action that requires an identification point for accommodating traction data, for example.

These numbering structures effectively service five areas of application where a particular type of identification is required, namely the identification of trade items, logistical units, assets, locations and service relations. A sixth area of application may be recognised in which standardised strings may be used for purposes internal to an organisation or for special applications not covered by the main application areas.

11.4.8 Data carriers and application identifiers – the EAN.UCC data standard

In addition to the numbering structures identified above the EAN.UCC system also distinguishes a number of barcode data carriers that have been adopted as EAN.UCC standards, supported also by an important set of EAN.UCC application identifiers. The capability of being able to use the numbering structures in data carriers that can also allow further data to be added and distinguished in a standardised way offers considerable flexibility in supporting item-attendant data handling and process improvement/innovation.

The data carriers adopted for EAN.UCC system applications presently comprise linear barcode symbols supported by the following standard symbology specifications:

- EAN/UPC symbologies including, UPC-A and UPC-E, EAN-13, EAN-8, and the 2- and 5-digit add-ons. These are symbologies specifically designed for omni-directional scanning at point-of-sale retail outlets and constitute the standard for use on items scanned in this way. The symbols may also be used on other trade items.
- Interleaved two-of-five (ITF-14) symbology for symbols carrying identification numbers on trade items not for scanning at retail outlets. The symbology is particularly suited for printing directly onto corrugated fibreboard and similar substrates.
- UCC/EAN-128 symbology, a particular variant of Code 128 exclusively licensed to EAN.UCC as the symbology supporting systems applications in which the system numbering and application identifiers are exploited. It is a variable length, alphanumeric symbology offering considerable flexibility for identifying and handling item-attendant data. The symbols are not intended to be scanned at point of retail, but within other areas of supply chain and industrial activity.

A wide and growing range of over 90, two-, three- and four-digit application identifiers (AIs) provides a framework for supporting the identification of application measures. AIs are also available to identify features such as logistics units expressed as a serial shipping container code (SSCC), serial numbers, batch and lot numbers, production and packaging dates to name but a few. A data format is specified for each AI to indicate the number and disposition of numeric and alphabetic characters. The AI to denote the identification of a logistic unit comprising the SSCC is 00, having the format n2 + n18, i.e. two digits (n) for the AI and 18 for the SSCC. An example of an alphanumeric AI is for a serial number (AI = 21) having the format n2 + an..20, denoting two digits for the AI and up to 20 alphanumeric (an) characters. The AIs for measures are grouped into metric and non-metric trade item measures and metric and non-metric logistic item measures for parameters such as weights, lengths, areas and volumes.

Where AIs are used in UCC/EAN data-carrier symbols, the AI precedes the data to be identified. All AIs for measurements are four digits long, the fourth digit being used to distinguish the position of the decimal point in the value data. For example, AI 330 is for gross weight of a logistics item expressed in kilograms and has the format n4 + n6. If the AI element is expressed 3302 and the data component has the value 002350 the 2 in the AI element places the decimal point between 3 and 5, so the weight is read as 23.50 kg. In the human readable form of the data encoded within carrier the AI is enclosed in brackets.

A reading device for EAN/UCC symbologies usually has the facility to generate, on reading a barcode symbol, a symbology identifier to be transmitted along with the element string as a means of distinguishing between the different EAN.UCC data structures and those of other barcode symbologies. Such facilities are important for achieving automatic processing of data, particularly for transactions and EDI message formatting. Having set the scene for embracing the potential of the EAN.UCC systems, further detail can be obtained from the GS1 UK website, www.gs1.org.

11.5   Chip-based data carrier technologies and radiofrequency identification (RFID)

Chip-based data carriers effectively exploit the capabilities and benefits of semiconductor memory technology, the type of memory usually associated with computers. The data carriers are essentially characterised by the way in which the memory devices are housed and accessed. Thus, one can distinguish card-based memory data carriers and tag-based carriers. An example of the latter is a memory device packaged in a form similar in construction to button-type batteries, hence the term button-memory often used to describe this type of carrier. The conductive surfaces of the can or package provide the conduit for data transfer to and from the memory chip. This accounts for the alternative term, 'touch-memory', often used to describe these devices. Data is generally contained in a non-volatile, static RAM memory (1–4 kbits) and transfer is achieved using a contact reader, allowing bi-directional transfer of digital data. Data integrity is provided through the use of cyclic redundancy coding (CRC), verification and page writes via an uninterruptable copy command.

Although the costs of touch memory data carriers are somewhat higher than label-based data carrier technologies, touch memory technology has the potential for providing cost-effective solutions for a variety of identification problems and opportunities. It can be used in a number of ways for enhancing data management operations, where contact to interrogate the device is acceptable and the need is seen for a read/write device that can store reasonable amounts of data and can tolerate reasonably harsh environments.

Applications include the tagging of containers, in situ and on conveyers, totes, pallets, component reels, carts, cargo containers and loading dock location of trailers. Novel structures can be effectively applied for automating the data gathering activity, on conveyors, forklift trucks and other material-carrying equipment, through the use of low-power electrical contacts that allow contact to be made automatically with one or a series of inter-linked touch memory devices. Touch memory devices may also be used for access control and other security applications.

11.5.1   Radiofrequency identification

Radiofrequency identification (RFID) is an important area of data carrier development, with new generation systems and products offering considerable potential for low-cost data carrier applications. RFID covers a range of data carrying technologies, for which the transfer of data from the data carrier to host is achieved via a 'radiofrequency' link. This contrasts with the touch memory type carriers in which the data transfer is via a conductive pathway.

Although referred to as radiofrequency, RFID systems data carrier frequencies range from relatively low frequency (<135 kHz), through megahertz frequencies to gigahertz (microwave) frequencies. The radiofrequency link allows non-line-of-sight reading of the data and, by virtue of their construction, the data carriers or transponders (tags) are capable of operating within a range of conditions unsuitable for linear barcodes, multi-row barcodes and matrix-based codes. Carriers can be read-only or read/write depending upon the construction, with data carrying capability again determined by device design. The range of communication and data transfer capabilities depend upon device designs and operating characteristics.

Unlike optical scanners the carrier frequencies (and power levels) at which RFID devices operate are subject to legislative constraints, with bands within the regions of 135 kHz, 13.56 MHz, 868–960 MHz and 2.45 GHz being favoured for RFID applications. At present, the only carrier frequencies that appear to have 'international' acceptability are <135 kHz, 2.54 GHz and possibly 13.56 MHz. However, there can be differing requirements in respect of power or field strength allowable and other spectrum allocation parameters according to country. Despite these constraints RFID is gaining attention within the user community and, with the introduction of international (ISO) standards, the issues of universality and interoperability are also being addressed.

A radiofrequency identification system comprises a set of transponders or tags, a reader/interrogator and a means of programming or writing to transponder (Fig. 11.16). For some types of tag the manufacturer undertakes programming. These are invariably referred to as read-only tags, but user-programmable read-only tags may also be distinguished. Tags are often classified as passive or active, being determined by how the tag derives its power. If powered directly from the read or interrogating field of the RFID system, the tag is considered passive, whereas if a battery powers the tag, for example, it is considered to be active. A category of tag may also be distinguished in which a battery is used to support functionality of the tag circuitry while also being dependent upon power received from a reader in order to effect a response. This category of tag is generally referred to as battery assisted.

In general, active tags allow a greater communication range than can be achieved for passive devices. Generally speaking they also offer better noise immunity and higher data transmissions rates when used to power a higher frequency response mode.

Passive tags are often constrained in their capacity to store data, the power output and corresponding range of communication and the ability to perform well in electromagnetically noisy environments. Sensitivity and orientation performance may also be constrained by the limitation on available power. Despite these limitations, passive tags offer advantages in terms of cost and longevity. They have an almost indefinite lifetime and are generally lower on price than active tags. A range of tags can be identified, varying in complexity. They may be viewed in terms of read/write capability, data

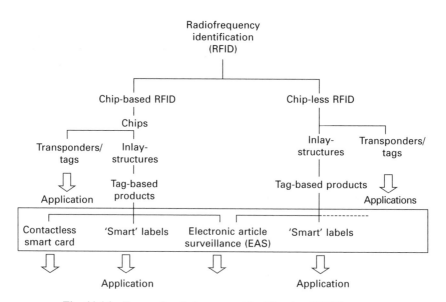

Fig. 11.16 Types of radiofrequency identification (RFID) system.

capacity and handling capability and anti-contention features (ability for reading multiple tags present in the interrogation or read zone at the same time). On the basis of data capacity, devices can be distinguished that range from what are essentially single bit electronic article surveillance (EAS) tags through 64 bit, 128 bit manufacturer/user programmable devices, user programmable 512 bit devices to devices to over 32 kilobytes of data storage capacity for upper range active tag devices.

The EAS devices are being used increasingly in retail outlets to deter theft, an alarm being activated when the device enters an interrogation zone. The lower-capacity tags are suitable for holding a serial or identification number together with parity check bits. Tags with data storage capacities up to 512 bits are suitable for accommodating identification and other specific data such as serial numbers, package content, key process instructions or possibly results of earlier interrogation/response transactions.

Devices having data storage capacities up to 64 kbits or above provide a portable data file capability, allowing useful amounts of data to be stored which have relevance to processes to which the tagged item may be subjected. In view of the higher data capacity, relative to other levels of complexity, the need can be seen for higher data transfer capability, particularly if a transponder is required to deliver all its data on interrogation and is moving through the interrogation field. In general terms, the higher the carrier frequency the higher the data transfer capability. However, it has to be appreciated that the data rate is determined by a number of factors including, significantly, the modulation method. Consequently, it does not always follow that higher-frequency carrier systems have higher rates than some lower-frequency carrier systems.

11.5.2   International standards

As far as supporting standards for RFID systems are concerned, international standards are now available, with others, including application standards, in prospect. International Standards Organization (ISO) standards have been available for some time for animal identification (ISO 11784 and 11785 with further development through ISO 14223/1) and contactless smart cards (ISO 10536, ISO 14443 and ISO 15693). Other standards, having a specific application focus, include identification for freight containers using 2.45 GHz transponders (ISO 10374).

The need to produce broader based standards to accommodate supply chain item management requirements has resulted in significant standardisation activity being pursued through ISO/IEC JTC1 SC31 WG4 - RFID Item Management and the publication of the ISO 18000 series air-interface and data-structure standards:

18000-1 Part 1 – Reference architecture and definition of parameters to be standardised

18000-2 Part 2 – Parameters for Air Interface Communications below 135 kHz

18000-3 Part 3 – Parameters for Air Interface Communications at 13.56 MHz
18000-4 Part 4 – Parameters for Air Interface Communications at 2.45 GHz
18000-5 Part 5 – Parameters for Air Interface Communications at 5.8 GHz – abandoned project.
18000-6 Part 6 – Parameters for Air Interface Communications at 860 to 930 MHz
18000-7 Part 7 – Parameters for Air Interface Communications at 433 MHz

A number of related standards include:

- ISO/IEC 15961:2004 Information technology – Radiofrequency identification (RFID) for item management – Data protocol: application interface
- ISO/IEC 15962:2004 Information technology – Radiofrequency identification (RFID) for item management – Data protocol: data encoding rules and logical memory functions
- ISO/IEC 15963:2004 Information technology – Radiofrequency identification for item management – Unique identification for RF tags

Further standards are in the process of being developed including conformance and application standards.

A very significant applications feature used in a number of the air interface standards is the application family identifier (AFI). This is an identifier facility that allows de-selection, at the earliest opportunity, of tags that are not relevant to a particular application. Used effectively AFIs allow the same air interface standard to be used for different applications without contention.

11.5.3   Electronic Product Code

A significant development with respect to numbering and RFID standardisation, particularly with supply chain applications in mind is the Electronic Product Code (EPC). EPC is essentially a numbering system, that can be related to legacy systems such as EAN.UCC, designed to allow unique identification down to individual item level. Whereas the EAN.UCC system is intrinsically capable of unique identification, the EPC facility can accommodate a greater set of numbers and can be supported by a numbering standard and a set of data carrier, network and infrastructure standards. The data carrier standard essentially relates to EPC-specified RFID data carriers and with a particular emphasis upon UHF carrier frequencies. High-frequency (13.56 MHz) and indeed other data carriers have also been indicated as possible support platforms for EPC. Alignment with ISO/IEC standardisation is being pursued with respect to an EPC specified UHF air interface and the use of AFIs (ISO/IEC 18000-6 Type C, relating to EPCglobal Class 1, Generation 2 tags).

The object of the EPC initiative is to facilitate very low-cost tags and an infrastructure that could accommodate the management of tagged items through the use of the EPC number and information concerning items held externally, accessed and functions handled through a network of access nodes and supporting architectural transfer, storage and processing facilities.

11.5.4   Applications

Within the food supply chain opportunities may be seen for applying RFID to palletised or container carriers of items for the purposes of identifying container contents on a regular basis and re-writable to allow differing container contents to be identified without recourse to changing labels. Reuse in this way will amortise the cost. The actual cost of the tags obviously depends on the type and quantities that are purchased. For large quantities (tens of thousands) of the lower-priced devices, the costs can range from less than a few tens of cents for extremely simple tags to tens of dollars for the larger and more sophisticated devices. Increasing complexity of circuit function, construction and memory capacity influence the costs of tags and reader/programmers.

The manner in which the tag is packaged to form a unit also has a bearing on cost. Some applications where harsh environments may be expected, such as abattoirs, will require mechanically robust, chemical and temperature tolerant packaging. Such packaging will invariably represent a significant proportion of the total transponder cost. Where very large-volume, low-cost tag applications are identified, such as packaging tagging, label and card structures are seen as appropriate.

A significant feature of new generation RFID tags is the ability to read a number of them when present in the interrogation or read zone at the same time. This feature is known as contention management whereby each tag can be read without the reader being overwhelmed or confused by the signal responses received. Various techniques have been developed to handle or avoid contention and it is important to be aware of the need when selecting tags. Such systems open up the opportunities for batch reading applications within the supply chain where, because of conditions or data capacity and transfer requirements, barcoding and other forms of identification would be inappropriate.

Applications where RFID is seen to have particular benefit include the tracking of totes, carriers, pallets and vehicles under conditions where non-contact exchange of data is required. RFID tags are unaffected by the grease and humidity or the non-metallic vagaries of food handling environments. The read/write capability of some RFID tags makes them suitable for applications where there is a need to alter data content on a fairly frequent basis. As the costs of tags reduce and standards are produced, applications for RFID identification will undoubtedly expand.

11.5.5   Electronic identification of livestock

There are three different types of RFID device, or electronic tags, used in the identification of livestock. These are electronic ear tags, injectable transponders and bolus tags, the choice being dependent upon type of animal and any legislative controls (see Chapter 9). The electronic ear tag is an RFID transponder enclosed in a protective plastic case that is attached externally to the ear of the animal. The tag is read-only and identifies the animal using a unique code. If the tag is lost or damaged, the animal can easily be re-tagged. The injectable transponder is an electronic tag that is injected under the cartilage in the animal's ear. The bolus tag is a cylindrical shaped electronic tag, or an injectable tag enclosed in a sealed cylindrical container. The tag is swallowed by the animal with the aid of a special bolus applicator and locates within the fore-stomach of the animal. Such devices are consequently only suitable for ruminant animals.

All the tags are read-only, but have specific advantages and disadvantages concerning ease of reading and application and likelihood of recovery. Electronic eartags are less secure as they are more easily lost, damaged or removed. However, they are much more straightforward to read, apply and recover. Bolus and injectable tags both require expertise and time in application and recovery and bolus, in particular, are more difficult to read than the eartag and injectable tags, which are easily accessible with a mobile reader at the ear. On the other hand, injectable and bolus tags are far more secure than the eartag, suffering few losses and damage, and protecting against unauthorised removal.

The tags that are used for animal identification are invariably low-frequency devices (<135 kHz) and generally conform to the ISO standards for animal identification, ISO 11784 (specifying data structure) and ISO 11785 (specifying the air interface). Although the data structure standard specifies a country code, together with a national identification code, a practical limitation arose in relation to the factory programming of these devices (read-only) and the delays that were being imposed between order and delivery or expense being incurred through stocking at country level to satisfy needs. This problem was addressed through the development of an additional standard (ISO 14223) that allows the use of read–write tags together with algorithms for proving authenticity of tags and prevention against illegal copying. The standard had to be compatible with ISO 11784 and a switch code is specified to allow mode selection between them.

11.5.6 Smart labels

Smart labels constitute an important area of passive RFID device development, offering the prospect of carrying both static and dynamic data, the latter by virtue of read/write capability. These are essentially short-range, low-cost (less than €1, in quantity) data carriers that emulate the function of label or, more prominently, relate to label structures in which a tag and associated antenna (an inlet device) are incorporated or laminated into the label form. Such labels allow for over printing. Such has been the impact of this technology that an AIM Industry Standard for 13.56 MHz RFID Smart Labels and RFID Printer/Encoders is being proposed. Most of the smart label products operate on a 13.56 MHz carrier frequency, but other smart label products are available that operate at UHF (862–928 MHz) and 2.45 GHz carrier frequencies.

The inlet (or inlay) concept supports the realisation of chip-based labels, in which the tag component is laminated between paper substrates to yield a label structure. The label can often be over printed with graphics, text or even printed data carriers such as linear barcodes, multi-row barcodes and matrix codes. In addition to inlet structures, 13.56 MHz discrete tags are also available to support a range of reuse applications. The advent of the ISO 15693 vicinity card standard has undoubtedly influenced the development of compliant chip-based devices for smart labels and ISO 15693 compliant products are now appearing that offer a range of capabilities including:

- Memory options, typically 2.5 kbits (32 pages, each of 8–10 bytes for user data and 2 bytes for function management) and 10 kbits (128 pages, each of 8–10 bytes for user data and 2 bytes for function management).
- Security functions, such as message and mutual authentication, based upon cryptographic algorithms with key lengths of typically 64 bits.
- Anti-contention capability for batch readability (compliant with ISO 15693).
- Electronic article surveillance (EAS) capability.

11.5.7 Smart active labels

The smart active label (SAL) may be viewed as an evolutionary extension of the chip-based smart label technology, but with added complementary attributes. A SAL may be considered as a structure in which a low-cost RFID inlet or inlay (tag comprising RFID chip and antenna structure) is connected to a low-profile power source and embedded into a label or similar construction capable of carrying printed information and graphics. Although they share some of the physical attributes of passive smart labels, the capabilities engendered through the availability of integral power provision distinguish the SAL devices quite clearly from their passive counterparts.

It is through the developments of low-profile and thin-film battery technology that SALs are now becoming a reality and, while cost is still an issue, the diversity of application needs and opportunities present a tangible growth market for such devices. The market will be driven by item-focused identification and management needs that can be fulfilled by SAL-based systems. From a supply chain perspective, for example, various levels of identification and item-management needs can be recognised, through:

- Packaged items – individual or mixed fundamental items in a single package.
- Transport units comprising a number or set of packages or other discernible items.
- Unit load or palletised unit – carrying a number of transport units or other discernible items.
- Container units – for accommodating pallets or other discernible items.

Thus, it is possible to recognise a hierarchy of item-related needs and distinguish technology appropriate to those needs. While SALs for individual items may be indicated on an item–label cost ratio basis, SALs may similarly be justified for other items such as packages, transport units, unit loads and even containers, depending upon the robustness and sophistication of the SAL. As costs reduce so the application arena can be expected to expand.

As with passive smart labels, SALs have the facility to carry both static and dynamic (changeable) data. The power capability of a SAL provides the facility to support functions that are not generally possible, or rudimentary supported, with passive devices. These functions include enhanced range capability, sensory, processing, display and locating capabilities. Enhanced security can also be seen as a further capability that can be suitably supported.

SALs, typically being designed to operate at UHF carrier frequencies of 433, 868 and 915 MHz. In all cases, the power levels are subject to national and or international regulatory constraints and spectrum allocation in terms of available bandwidth and other spectrum usage features such as duty cycle. Active RFID systems operating at 433 MHz carrier frequency represents the most widely used and, globally, the most acceptable carrier frequency for active RFID.

11.6   The electronic product code (EPC) system

The electronic product code (EPC) system is an item numbering and networking concept that emerged from research undertaken at the Massachusetts Institute of Technology (MIT) Auto-ID Center, and associated centres, in which RFID data carriers were identified as the method of choice for carrying the EPC numbers. The original concept sought to specify a 96-bit number structure that could be used to uniquely identify individual items and facilitate the management and associated uses of those items through 'intelligent' (savant) terminals and information stored elsewhere. The numbering and associated data carriers, together with the exploitation of available connectivity, were seen as the basis for an 'Internet of Things' or a 'Networked Physical World'.

Developments towards the realisation of this concept were sponsored and undertaken with the support and involvement of the Uniform Code Council (UCC) and a consortium of potential user organisations such as consumer goods manufacturers, retailers and supply chain logistics providers. The consortium also included RFID technology developers. In November 2003, the responsibility for the commercialisation and management of the EPC system was transferred to EPCglobal Inc. EPCglobal Inc. is affiliated to UCC and to EAN International, both of which support and promote an internationally recognised identification and numbering scheme for consumer goods management around the world. As with the EAN.UCC numbering and identification scheme, which is supported by a number of EAN and UCC adopted barcode symbologies, the EPC also represents a coding scheme to support item management. Logically this should embrace the legacy of the EAN.UCC system and, while this was not evident in the initial formulation, subsequent developments are in place to accommodate this requirement.

Indeed, a migration strategy is now being promoted in which the existing Global Trade Item Number (GTIN), currently used for product identification within the EAN.UCC numbering and identification scheme using barcode data carriers, is being adopted within the EPC coding structure.

Developments of the EPC concept and system specification have been attended by a great deal of hype and mis-information, but now appear to be geared to fulfilling its defined objectives. In this progression other needs have been identified and accommodated in the EPC systems specification, including a range of RFID-based data carriers to support a range of additional needs. The EPC system, as expressed in its original conceptualisation, comprises a number of parts, as follows.

11.6.1 EPC numbering scheme

This initially comprises a 96-bit code structured as an 8-bit header, followed by three data partitions for EPC Manager (28-bits to facilitate identification of the item manufacturer or source provider for example), Object Class identifier (24-bits to facilitate the identification of the type of product, such as a specific stock-keeping-unit) and serial number (36-bits to facilitate the unique identification of an individual item). The header, also known as the EPC version number, is used to distinguish multiple EPC formats, thus allowing designation of differing bit-lengths tags as the technology matures (a 256-bit version is in prospect). The header can also be used to distinguish bit-length field variations to those indicated above with respect to manufacturer, product and serial number support, so allowing longer and more manufacturer identifiers for organisations with a small number of product types and serial number requirements.

While the initial bit-length designation for EPC was 96 bits, a shorter 64-bit version has been introduced on an interim basis to help facilitate the realisation of lower cost RFID data carrier devices. This is reasoned on the basis that the full identification capability of the 96-bit version would not be required for some time.

11.6.2 EPC object naming service (ONS)

This was introduced to provide a 'directory service' capable of supporting the linkage of EPC numbers with additional data or information concerning the item to which the EPC tag is attached. This additional, item-associated data or information may be stored on a server connected to a local network or the Internet. The ONS is analogous to the Domain Name service used for location of information on the Internet.

11.6.3 EPC physical markup language (PML)

This was structured to allow the information about an item or object to be appropriately specified. The PML is based upon the popular XML meta-data language, its syntax and semantics to be administered and developed by the governing body (EPCglobal) in conjunction with the user community. Product definitions within this language markup facility, which commenced with food items, require the on-going efforts of the governing body to build a sufficiently embracing directory. Product descriptions already undertaken by standards bodies such as International Bureau of Weights and Measures and the National Institute of Standards and Technology, are being seen as valuable sources of information in this respect.

In addition to fixed product information the PML will also accommodate dynamic quantities, such as temperature, humidity or vibration, which may change as a result of some locational, environmental or intrinsic effect, including changes over time (temporal effects). This adds further dimension to the data gathering and handling processes and, when presented in a PML file, may offer innovative opportunities for process enhancement. For example, condition status information derived dynamically could be used to automatically determine product pricing.

11.6.4 EPC Savant software

This software is being developed to facilitate a broad and ambitious set of functions to manage item data and information. Its most basic function is to receive the EPC number and direct a query over the internet or other established network to the ONS which then returns an address at which the item information is stored. The information is available to, and can be augmented by, Savant systems within the network, ostensibly around the world. The very high data handling envisaged within the EPC infrastructure indicates the potential need for companies to maintain ONS servers locally to support rapid retrieval of information.

The Savant software is being developed to use a distributed architecture with a hierarchical structure to manage data flow. A highly extensive network of Savants is envisaged to support the EPC data management, with Savant platforms running, for example, in factories, stores, distribution centres, regional support facilities and even on mobile platforms such as container lorries and cargo planes. Creating such an infrastructure is one of the biggest challenges to realising the EPC support objectives. Added to this is the challenge of accommodating a wide range of Savant functions, including:

- **Data smoothing** – to receive data from EPC readers and effect particular filter (data smoothing) functions and error control (error detection and correction) before transfer.
- **Reader coordination** – to manage signals from different readers and delete duplicate reads arising from the same tag being read by two or more readers at the same time.
- **Data forwarding** – wherein the Savant is required to decide which information needs to be forwarded up or down the chain in accordance with application needs.

- **Data storage** – to maintain a real-time in-memory event database (RIED) for the purpose of avoiding database overloads that may otherwise occur due to the very high transaction rates expected in EPC systems usage. The Savant effectively takes the EPC data that is generated in real time and stores it 'intelligently' in order to allow other enterprise applications access to the information without causing database overloads.
- **Task management** – in the form of a task management system (TMS) that is integral to all Savants regardless of hierarchy and geared to facilitating data management and monitoring tasks on an individual application basis. For example a Savant running in a distribution centre may alert a distribution manager to a possible cross-docking difficulty due to a delayed arrival of goods and possibly alert the manager to an alternative solution.

11.6.5   EPC specified data carriers

The data carriers represent the means by which EPC numbers are carried and attached to items to be identified. Strictly speaking the data carriers can be of any form providing they are capable of carrying the EPC numbers, can communicate effectively with EPC readers and satisfy EPC-specified performance requirements. The original EPC brief perceived electromagnetic identification devices (EMIDs) for this purpose and moved towards the specification of RFID (high and ultra-high frequency) air interface requirements for communication and data transfer, without otherwise imposing constraints on manufacturer tag and system designs.

The various categories of data carrier that can now be distinguished include:

- **Class 0 tags** – read-only, 64- or 96-bit memory devices designed to operate at UHF carrier frequencies in bands between 868 and 930 MHz to accommodate differing regulatory requirements in different countries. A read–write variant Class 0+, sharing the same technical protocol as Class 0, is offered by Matrics Inc.
- **Class 1 UHF tags** – one-time programmable (OTP) devices designed to operate at UHF carrier frequencies in bands between 868 and 930 MHz with up to 96 bits of memory. An EPC number is encoded into the tag on manufacture, the remaining 96 bits of memory being available to the user to programme as appropriate for the application intended, whence the tag becomes read-only, with both EPC number and memory content being available for use.
- **Class 1 HF tags** – one-time programmable (OTP) devices designed to operate at a high frequency carrier of 13.56 MHz.
- **UHF Generation 2 (Gen2) specification** – developed to overcome concerns over interference and regulatory variations concerning allowable carrier frequencies in different parts of the world. The Gen2 specification extends the memory capability from 96 bits to 256 bits. As with the class 0 and class 1 UHF tags, the Gen2 tags are designed to operate at carrier frequencies within the 860–930 MHz band and in compliance with the ISO 18000-6

air–interface standard for UHF RFID data carrier systems – ISO/IEC 18000-6 Type C development.

- **Class 2 read/write tags** – devices designed to support EPC numbers, together with additional data and read/write capability for applications requiring item-attendant data changes over a period of time.
- **Class 3 read/write tags with sensory capability** – devices designed to support EPC numbers, together with facilities for acquiring, storing and communicating sensory data, such as temperature, humidity and vibration.
- **Class 4 active read/write tags** – devices designed with onboard battery power to support tag functionality, including integral on-board power supported data transmission.

In hardware terms, the essential elements of the EPC systems comprise:
- Tag-based data carriers
- EPC compliant readers
- Savant, middleware holding systems
- ONS database systems
- PML Servers.

These elements connect to form the data exchange spurs of the EPC infrastructure or network (Fig. 11.17).

Despite the substantial investment and effort that has gone into the development of EPC since its inception there is still a significant amount of effort required to establish the envisioned infrastructure for EPC and the associated challenges in realising the Savant functions. Nevertheless the pilot studies and support given by giants such as Wal-Mart in driving the initiative suggest that it will become a reality. The developments have certainly

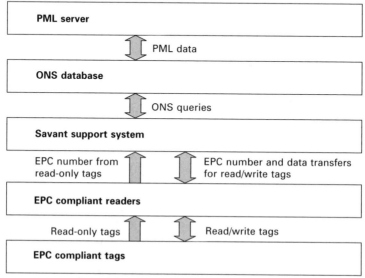

Fig. 11.17 EPC infrastructure.

had a significant impact upon the design and manufacture of low-cost tags, an important requirement of the EPC vision. The developments have also raised significant issues, not least of which are the largely unfounded concerns over privacy, issues concerning alignment of standards with relevant ISO standards and the issues concerning harmonisation of regulatory frequencies to support global use of EPC adopted RFID data carriers. All of these issues are being tackled and in the fullness of time will undoubtedly be resolved.

As far as usage of the EPC numbering system is concerned this is seen as requiring membership of the EPC global body with associated costs for acquiring numbers. This is similar in effect to the EAN and UCC membership requirements for using the EAN.UCC system for numbering and identification. For further information and developments on EPC see http:// www.epcglobalinc.org/.

11.7 Summary

A wide range of data carrier technologies and associated products are available for supporting traceability at the various levels of item identification. The more prominent of these technologies are summarised in Fig. 11.18. Many but not all are supported by international standards. In order to support full open system traceability, further standardisation is needed and this should be born in mind when considering the choice of data carrier and data structures for secondary identification and process support.

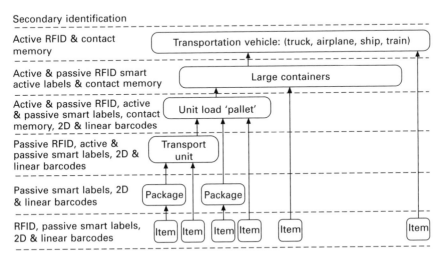

Fig. 11.18 The range of data carrier technologies.

Index